The
Pollution Control Policy
of the
European Communities

The Pollution Control Policy of the European Communities

Stanley P. Johnson
Vice-Chairman, European Parliament's
Committee on Environment, Public Health
and Consumer Protection

Graham & Trotman

First published in 1983 by
Graham & Trotman Limited
66 Wilton Road
London SW1V 1DE

© Stanley P. Johnson, 1983

British Library Cataloguing in Publication Data
Johnson, Stanley, *1940—* *The pollution control policy of the European* *communities.*—2nd ed. 1. Environmental policy—European Economic Community countries. 2. European communities I. Title 363.7′3′094   HC260.E5 ISBN 0 86010 411 7

ISBN 0 86010 411 7

Typeset in Great Britain by Input Typesetting Ltd, London
Printed and bound in Great Britain by the Camelot Press Limited,
Southampton

Contents

Tables

ix

'The Community is designed to protect and advance the interests of *all* its citizens. Policies to safeguard the producer need to be balanced by policies to safeguard the consumer. That balance has not always been struck in the past. This means that we should, I think, give greater weight to the protection of the consumer as well as to that of the environment in which the consumer—you and I—live.'

Address by the Rt. Hon. Roy Jenkins, President of the Commission of the European Communities to the European Parliament, Luxembourg, 11 January 1977.

'What, you will ask, should the European Union's area of responsibility be? Firstly, to my mind they must include the powers which the Community currently exercises or could exercise. Some Community policies are expressly provided for in the Treaties, while others—the so-called grey areas—have been or remain to be evolved on the basis of general powers, such as those envisaged in Articles 100 and 235 of the Treaty. Some activities have already been initiated on this basis, such as the regional and environment policies. Others still in gestation include monetary policy. It is essential that these various aspects of Community action should be accorded definite European status, to put them beyond dispute like, for example, the Common Agricultural Policy, whose status they should ultimately share.'

Statement by Gaston Thorn, President of the Commission of the European Parliament at a meeting of the Parliament's Institutional Affairs Committee on 29 April 1982

'The European Council underlined the urgent necessity of accelerating and reinforcing action at national, Community and international level aimed at combating the pollution of the environment'.

Declarations and Conclusions of the European Council, Stuttgart, 20 June 1983.

Preface to Second Edition

This book was first published in 1979. Since then the European Community's activities in the field of environment have continued to be vigorously prosecuted. Today over 80 pieces of legislation have been adopted, the vast majority of them relating to pollution prevention and control.

The second edition of this book, like the first, concentrates on the pollution control policy of the EEC. That, after all, is the book's title. It must be recognized however that certain major developments which are now taking place, such as the discussion of Environmental Impact Assessment and the Environmental Fund, have an importance which goes far beyond questions of pollution. If my treatment of such items is limited, this is because the scope of the book itself is limited.

These last four years have seen the increasing involvement of the European Parliament in environmental questions. The European Parliament, in particular, has pressed for the creation of the European Environment Fund, to take its place alongside the Regional Fund, the Social Fund, etc. The Parliament has also been concerned to see the Community itself take a more responsible attitude towards projects or programmes—e.g. agricultural drainage schemes—which are themselves being financed out of Community resources and which may have undesirable effects, including the creation of pollution. At Community level, as is often the case at national level, the right hand seems often not to know, or not to care, what the left hand is doing.

As indicated by the quotation from Commission President Gaston Thorn which is given at the front of this book, an attempt is now being made to draw up a new EEC Treaty more relevant to the needs of the present and future. Whether or not this effort bears fruit, there can be no doubt that environmental policy has over the last few years moved into the forefront of the European Community's concerns. As I write this note, the results of the West-German election—with twenty-seven seats going to the Greens—have just been made known. 'Green' parties, of whatever sort, are the political manifestation of the more general proposition that 'the environment' has at last come of age.

I am immensely grateful for Eric Lummis for his assistance in preparing this revised edition.

Stanley P. Johnson
Strasbourg, March 7, 1983

Preface to First Edition

The policy of the European Communities on the control of pollution is not a matter of opinion or conjecture. It is not a 'grey zone', situated at or even beyond the margin of Europe's legitimate concerns. On the contrary, the European Communities' pollution control policy is clearly defined by a collection of instruments which in one form or another have been explicitly noted, approved or adopted by the Community institutions.

At the most general level, the European Communities' policy on pollution control is contained in the Declaration of the Council of the European Communities and of the Representatives of the Governments of the Member States meeting in the Council of 22 November 1973, on the programme of action of the European Communities on the Environment, and in the Resolution of the Council of the European Communities and of the Representatives of the Governments of the Member States meeting within the Council of 17 May 1977 on the continuation and implementation of a European Community policy and action programme on the environment.

In addition to these two basic documents (large parts of which are devoted to pollution control), we have to consider a number of specific instruments which, having been adopted by the Council, must be seen as filling out in more detail various aspects of the general policy. This volume describes the texts which have been already adopted by the Council in the various fields of pollution control—water, air, waste, noise, chemicals, and so on. It also describes those proposals for Community action which have been approved by the Commission even though they have not yet been finally approved by the Council. (In the field of environment, as in other fields covered by the Treaty of Rome, the Commission as general rule 'proposes' and the Council 'disposes'.)

The approach taken, therefore, is purely documentary. It is an attempt to assemble a collection of texts—some well known, some less known—into a coherent whole. The facts which are cited, the motives and methods which are described and the details which are given of a particular proposal or project, are those officially approved and published by the various Community institutions. There is nothing secret or confidential here. No inside material has been used.

Because this has been an evolving and dynamic programme, the approach followed in one legal instrument, such as a Directive, may be modified or expanded in a subsequent one. As far as the environment is concerned, the Community has a relatively short history but even five years is long enough for certain lessons to be drawn from the past and for new approaches to be defined for the future. Where mention is made, therefore, of a Commission or a Council decision on some particular topic, care has always been taken to 'datestamp' that decision in a precise fashion.

Fundamental questions are, of course, still being asked and answered about the Community's environment programme: What is its legal basis? Article 100 or 235 or both? Or is the fact that the Council has adopted an environment programme in itself a sufficient legal basis regardless of specific articles in the Treaty? Should the attempt to reduce distortions in the conditions of competition also cover environmental factors? Are short, fast-flowing rivers, which rapidly carry waste to the sea to be considered a legitimate element in the basket of the factors of production on a par with, say, Mediterranean sunshine? What is the role of absorptive capacity? Is there a case for uniform emission standards?

It is not the purpose of this introduction to deal with these questions head-on. To summarize is to distort. But the reader, after studying the relevant texts, which are described at some length in the body of the book, should be able to form his own opinion and to avoid some of the 'snap' characterizations of the Communities' environment policy which have led to difficulties in the past.

The fact that this book is limited to the questions of pollution control in no way implies that pollution is necessarily the Communities' main priority in the field of environment. Steps are being taken in other areas, such as conservation and land-use planning which in the long run may be just as important as far as the 'quality of life' is concerned as the attempt to produce clear air and clean water. The fact is, however, that the vast bulk of the legislation which has so far been adopted at Community level is concerned with pollution. Ten years from now it may be a different story and there may be a different set of texts to bring together.

No introduction to a book dealing with the Community's environmental policy would be complete without a heart-felt acknowledgement of the leadership and vision of Mr. Michel Carpentier, the first Director-General of the Environment and Consumer Protection Service of the Commission of European Communities. More than anyone else he fashioned the policy and built up the resources which were necessary to carry out the Commission's part of the programme. Indeed the work of all his staff is acknowledged here, and not least that of Miss Heather Sowerby who has worked with patience and good humour on this project.

The effort has, of course, been collective in the broadest sense, for it is not the Commission's environment policy. It is the Community's environment policy. Which means that the commitment of human and material resources has had to be made by the institutions of the Community and the

Member States acting together. If this book needs any dedication, then it must be dedicated to all those who have been involved in the fight for a better environment—whether it be in Brussels or in the various capitals or in the often unsung and unvisited regions of Europe where policies, once decided, have actually to be put into effect.

Stanley P. Johnson
Brussels, 4 January 1979

1

Objectives, Principles, General Description of Environmental Programme

1. BACKGROUND

Among the objectives set out in the preamble to the Treaty establishing the European Economic Community, the signatories spoke of the need to achieve 'the constant improvement of the living and working conditions of their peoples' and 'the harmonious development of their economies'.

In Article 2 of the Treaty, the following are included in the statement of the task assigned to the Community; to promote throughout the Community a harmonious development of economic activities, a continuous and balanced expansion, and accelerated raising of the standard of living and closer relations between the States belonging to it.

When they met in Paris on 19 to 20 October 1972, the Heads of State or of Government of the Member States declared that 'economic expansion is not an end in itself: its first aim should be to enable disparities in living conditions to be reduced. It must take place with the participation of all the social partners. It should result in an improvement in the quality of life as well as in standards of living. As befits the genius of Europe, particular attention will be given to intangible values and to protecting the environment so that progress may really be put at the service of mankind'.

This readiness to ensure that the Communities direct their activities towards the improvement not only of the standard of living but of living conditions and the quality of life is expressed still more precisely in point 8 of the final Declaration of the Paris Summit Conference: 'the Heads of State or of Government emphasized the importance of a Community environmental policy. To this end they invited the Community institutions to establish before 31 July 1973, a programme of action accompanied by a precise timetable'.

The programme of action of the European Communities on the environment was established in response to this invitation. It also takes into account

1

the results of the Conference of the Ministers responsible for environment questions held in Bonn on 31 October 1972, together with the memoranda and documents forwarded by the Member States and the detailed comparison of the points of view of the Commission and the representatives of the Member States on the Commission's communication forwarded to the Council on 24 March 1972. It also takes into account the opinions expressed by the European Parliament, the Economic and Social Committee and the employers' and workers' organizations.

On 22 November 1973, the Council of the European Communities:

- approved the principles and objectives of a Community environment policy and the general description of the projects to be undertaken at Community level as defined in the programme;
- approved the material content of the projects in the programme for the reduction of pollution and nuisances and the priorities established in this matter;
- approved the specific guidelines of the projects in the programme for the improvement of the environment;
- noted that the Commission, as far as it was concerned, would be responsible for the implementation of the programme in accordance with the procedure and timetables indicated therein and that the latter would subsequently put forward suitable proposals;
- undertook to decide on the proposals mentioned above within a period of nine months from the date of their despatch.

2. OBJECTIVES OF COMMUNITY ENVIRONMENT POLICY

The aim of a Community environment policy as set out in the Council Declaration of 22 November 1973 is to improve the setting and quality of life, and the surroundings and living conditions of the peoples of the Community. It must help to bring expansion into the service of man by procuring for him an environment providing the best conditions of life, and reconcile this expansion with the increasingly imperative need to preserve the natural environment.

It should:

- prevent, reduce and as far as possible eliminate pollution and nuisances;
- maintain a satisfactory ecological balance and ensure the protection of the biosphere;
- ensure the sound management of and avoid any exploitation of resources or of nature which cause significant damage to the ecological balance;

- guide development in accordance with quality requirements, especially by improving working conditions and the settings of life;
- ensure that more account is taken of environmental aspects in town planning and land use;
- seek common solutions to environment problems with States outside the Community, particularly in international organizations.

3. PRINCIPLES OF A COMMUNITY ENVIRONMENT POLICY

In its Declaration of 22 November 1973, the Council endorsed the general principles of a Community environment policy as worked out by the Ministers of the Environment at their meeting in Bonn on 31 October 1972.

These general principles, which were defined in the light of subsequent ideas and exchanges of views, are as follows:

(a) The best environment policy consists in preventing the creation of pollution or nuisances at source, rather than subsequently trying to counteract their effects. To this end, technical progress must be conceived and devised so as to take into account the concern for protection of the environment and for the improvement of the quality of life at the lowest cost to the community. This environment policy can and must be compatible with economic and social development. This also applies to technical progress.

(b) Effects on the environment should be taken into account at the earliest possible stage in all the technical planning and decision-making processes.

The environment cannot be considered as external surroundings by which man is harassed and assailed; it must be considered as an essential factor in the organization and promotion of human progress. It is therefore necessary to evaluate the effects on the quality of life and on the natural environment of any measure that is adopted or contemplated at national or Community level and which is liable to affect these factors.

(c) Any exploitation of natural resources or of a nature which causes significant damage to the ecological balance must be avoided.

The natural environment has only limited resources; it can only absorb pollution and neutralize its harmful effects to a limited extent. It represents an asset which can be used, but not abused, and which should be managed in the best possible way.

(d) The standard of scientific and technological knowledge in the Community should be improved with a view to taking effective action to conserve and improve the environment and to combat pollution and nuisances. Research in this field should therefore be encouraged.

(e) The cost of preventing and eliminating nuisances must, in principle, be borne by the polluter. However, there may be certain exceptions and special arrangements, in particular for transitional periods, provided that they cause no significant distortion to international trade and investment. Without prejudice to the application of the provisions of the Treaties, this principle should be stated explicitly and the arrangements for its application, including the exceptions thereto, should be defined at Community level. Where exceptions are made, the need to progressively eliminate regional imbalances in the Community should also be taken into account.

(f) In accordance with the Declaration of the United Nations Conference on the Human Environment adopted in Stockholm, care should be taken to ensure that activities carried out in one state do not cause any degradation of the environment in another state.

(g) The Community and its Member States must take into account in their environment policy the interests of the developing countries, and must in particular examine any repercussions of the measures contemplated under that policy on the economic development of such countries and on trade with them with a view to preventing or reducing any adverse consequences as far as possible

(h). The effectiveness of effort aimed at promoting global environmental research and policy will be increased by a clearly defined long-term concept of a European environmental policy.

In the spirit of the Declaration of Heads of State or Government at Paris, the Community and the Member States must make their voices heard in the international organizations dealing with aspects of the environment and must make an original contribution in these organizations, with the authority which a common point of view confers on them.

(j) The protection of the environment is a matter for all in the Community, who should therefore be made aware of its importance. The success of an environment policy presupposes that all categories of the population and all the social forces of the Community help to protect and improve the environment. This means that at all levels continuous and detailed educational activity should take place in order that the entire Community may become aware of the problem and assume its responsibilities in full towards the generations to come.

(k) In each different category of pollution, it is necessary to establish the level of action (local, regional, national, Community, international) that, as befits the type of pollution and the geographical zone to be protected, should be sought.

Actions which are likely to be the most effective at Community level should be concentrated at that level; priorities should be determined with special care.

(l) Major aspects of environmental policy in individual countries must no longer be planned and implemented in isolation. On the basis of a common long-term concept, national programmes in these fields should be coordinated and national policies should be harmonized within the Community. Such policies should have as their aim the improvement of the quality of life. Therefore, economic growth should not be viewed from purely quantitative aspects.

Such coordination and harmonization should in particular make it possible to increase the efficiency of action carried out at the various levels to protect and improve the environment in the Community, taking into account the regional differences existing in the Community and the requirements for the satisfactory operation of the common market.

This Community environment policy is aimed, as far as possible, at the coordinated and harmonized progress of national policies without however hampering potential or actual progress at the national level. However, the latter should be carried out in such a way as does not jeopardize the satisfactory operation of the common market.

Coordination and harmonization of this nature would be achieved in particular:

● by the application of the appropriate provisions of the treaties;
● by the implementation of the action described in the programme;
● by the implementation of the environment information procedure.*

4. GENERAL DESCRIPTION OF THE PROJECTS TO BE UNDERTAKEN UNDER THE PROGRAMME OF ACTION OF THE COMMUNITIES ON THE ENVIRONMENT

The Council, in its Declaration of 22 November 1973, recognized that the protection of the natural environment and the improvement of living conditions necessitated the implementation of actions of various kinds.

This involved on the one hand, the adoption of measures to reduce pollution and nuisances, and on the other, ensuring that the improvement of living conditions and ecological factors, which must now be considered as inseparable from the organization and promotion of human progress, be integrated in devising and implementing common policies.

For the Community and its Member States, this also involved carrying out certain projects within international organizations so as to avoid unnecessary duplication, cooperating with third countries and ensuring that

* See Annex I, (a) and (b).

the specific interests of the Community were taken into consideration by these organizations.

For these reasons the programme of action of the Communities on the environment included three categories of action:

1. Action to reduce and prevent pollution and nuisances.
2. Action to improve the environment and the setting of life.
3. Community action or, where applicable, common action by the Member States in international organizations dealing with the environment.

The Council also recognized that environment policy would have to be pursued in the context of other sectoral policies (social affairs, agriculture, regional policy, industrial policy, energy policy, etc.)

This book, as its title indicates, is primarily concerned with the first of these three categories, namely action to reduce and prevent pollution and nuisances.

5. ACTION TO REDUCE POLLUTION AND NUISANCES

(a) The Community Environment Action Programme of 22 November 1973 stated that specific measures to protect man and his environment against the pollution and nuisances which assail him, should be supported by an objective analysis of the facts and the results of studies which show up the various consequences, in particular in the ecological and the economic field, of the choice of any one of several possible measures.

Study of the problems raised in the fight against pollution revealed the existence of numerous gaps: gaps in scientific knowledge and methods of analysis and measurement, gaps in economic experience, especially as regards the cost of the damage caused by pollution and of the measures to counter this and, finally, gaps in statistical data.

The action programme called for a series of projects to be undertaken at Community level in order to provide a common basis for the evaluation of data and a common framework of methods and references. Such work would enable the measures indicated to be carried out: it would also avoid costly duplications and the adoption by Member States of various measures liable to create economic or social distortions within the Community.

The following tasks would have to be undertaken:

1. The laying-down of scientific criteria* for the degree of harm of the principal forms of air and water pollution and for noise. This action must go hand in hand with the standardization or alignment of the

---

* For the definition of this and other terms contained in this Programme see Annex II.

methods and instruments used in measuring these pollutants and nuisances. In the laying-down of criteria, priority would be given to the following pollutants: lead and lead compounds, organic halogen compounds, sulphur compounds and particles in suspension, nitrogen oxides, carbon monoxide, mercury, phenols and hydrocarbons.

2. The definition, on the basis of a common methodology, of parameters and the decision-taking process in connection with the laying-down of quality objectives.

3. The organization and promotion of exchanges of technical information between the regional and national pollution surveillance and monitoring networks. In due course this action would facilitate the implementation of a Community information system dealing with the data acquired by these networks and the inclusion of these in the world monitoring system envisaged by the UNO.

4. The adoption of a common method of estimating the cost of anti-pollution measures. During an initial stage an attempt would be made, in collaboration with the OECD, to establish methods of estimating the cost of air and water pollution, and the cost of countering pollution caused by certain industrial activities. This work would be rounded off by an analysis of the economic instruments which can be used under an environmental policy allowing for the application of the principle of making the polluter pay, without prejudice to the rules of the common market.

A study would also be made of the methods of estimating the cost to society of the damage to the environment with a particular view towards including these costs in a suitable form in national accounting figures in the determination of the gross national product.

Finally, a common method of classifying and describing anti-pollution activites would be developed.

(b) The environment action programme recognized, however, that anti-pollution policy could not be limited to the type of action described. Its essential aim should be the adoption of measures by the Community and Member States for the protection of the environment by reconciling that objective with the satisfactory operation of the common market.

At Community level, the following action would have to be taken:

1. The standardization or harmonization of the methods and techniques for sampling, analysis and measurement of pollutants. Priority would be given to the standardization of measuring methods for oils and natural gases having known or probable carcinogenic effects, photochemical oxidants, asbestos and vanadium.

2. The preparation of a list of quality objectives determining the various requirements an environment must meet, bearing in mind its allotted purpose. Community action would also be oriented towards the

search for long-term quality criteria with which the various parts of the Community environment would have to comply.

3. The determination of standards which, in certain cases, could be provisional, and which in the first instance would be concerned mainly with water pollutants.

4. The harmonization of the specifications of polluting products. In order to ensure effective protection of man and his surroundings, this harmonization, which was already being implemented in the elimination of technical barriers to trade, should be accompanied by studies on the noxious effects of pollutants contained in such products, the possibilities of changing their composition and, if necessary, their replacement by non-polluting or less polluting substitute products. Moreover, as far as is necessary common measures relating to the conditions of approval and inspection of the use of such products should be examined and implemented. Priority would be given to vehicles, noisy products and equipment, pneumatic drills, motor and other fuels and combustibles, cleaning and washing products.

5. Studies in individual industries on pollution caused by industrial activities and energy production, relating to the principal polluting industrial activities, carried out in cooperation with the competent authorities of the Member States and the industries concerned. These studies should permit the exact nature of the pollution problems to be established, the best technical and economic solutions to be found and if necessary, allow any aids to be standardized and a study to be made of the possibility of harmonizing principles or sets of other measures as regards certain industrial sectors.

   In the first phase, work would be undertaken on the following industrial sectors: paper and pulp, iron and steel and titanium dioxide manufacturing.

6. With regard to the problems raised by toxic or persistent waste, it would be necessary to pool thought and experience in order to assess the technical and economic aspects of the various possible means of action for eliminating such waste and to determine on that basis the measures to be introduced at Community level, e.g. harmonization of regulations, promoting of the development of new techniques, possible establishment of a system for pooling information, etc.

   Priority would be given to dangerous substances listed in Annex I, to the Oslo Convention, and to waste oils.

7. To avoid distortion of trade and investment, and without prejudice to the application of treaty provisions, the 'polluter pays' principle would need to be worked out and its terms of implementation, including the exceptions thereto, laid down at Community level.

8. Finally the serious problems posed by the pollution of certain zones of common interest (marine pollution, pollution of the Rhine basin and certain frontier zones) would require the introduction of special measures and procedures in a suitable framework, taking into account the geographical characteristics of such zones.

Thus, as far as marine pollution is concerned, Community action would consist in particular of:

- harmonizing the rules for implementing international conventions insofar as is necessary for the proper operation of the common market and the execution of the programme;
- implementing projects to combat land-based marine pollution along the coastline of the Community.

As regards the steps or the position to be taken during the work, Member States would endeavour to adopt a common attitude within the international organizations and conferences concerned, without prejudice to Community actions in respect of subjects falling under its competence or without prejudice to joint action with Member States might take, in respect of all matters of particular interest to the common market, within the framework of international organizations of an economic character.

With regard to the protection of the Rhine against pollution, the Commission was taking part as an observer in the plenary sessions of the International Commission for the Protection of the Rhine against Pollution. Moreover, in recalling the suggestions it made in its second Communication to the Council on the environment, the Commission reserved the right to make suitable proposals by 31 March 1974, taking into account the studies already started and the results of work in hand within the International Commission for Protection of the Rhine against Pollution following the Ministerial Conference at the Hague.

With regard to the protection of the environment in frontier zones, the Council recommended the Member States to establish consultation procedures for the conclusion of agreements on the protection of the environment in such zones.

9. Finally, the Council stated that common action on the environment implied that if compliance with Community or national regulations was to be effectively controlled, infringements against these regulations would be dealt with severely. To this end, the Commission would continue with its work of comparison of national laws and their application in practice in order to create the conditions necessary for the approximation of laws which prove necessary, and exchanges of information on the actual controls and the measures taken by each Member State would also be organized so as to ensure proper observance of the rules relating to polluting installations and products.

The measures referred to would be backed up by the implementation of a common research programme and by planning the establishment of a European system of documentation for the processing and dissemination of information on protection of the environment beginning with information on anti-pollution techniques and technologies and with the effects of pollution on human health and the natural environment.

With regard to the research programme, the Council noted that work on the environment already featured in both the Community Joint Research Centre's multi-annual programme and in the programme of so-called indirect research activities.

These research activities would have to be carried out, however, as a complement to the activities contained in the programme of action.

6. THE SECOND ENVIRONMENT ACTION PROGRAMME

On 13 June 1977 the Council of the European Communities and of the Representatives of the Governments of the Member States meeting within the Council of 27 May 1977 adopted a Resolution on the continuation and implementation of a European Community policy and action programme on the environment covering the period 1977 to 1981. The essential elements of this Second Environment Action Programme, in so far as they concern the EEC's approach to pollution control, are given in the individual sections of this book. The Resolution of 13 June 1977 restated in their entirety the objectives and principles of a Community Environment Policy as set out in the programme of 22 November 1973. It also recalled that, when adopting the 1973 action programme, the Council noted the determination of the Member States to ensure that the present quality of the various environmental areas, taking into account the Community regions as a whole, would not deteriorate, in view of the often irreversible or practically irreversible nature of some pollution. The Commission would continue the studies it had begun on ways and means of making this approach a reality.

7. THE THIRD ENVIRONMENT ACTION PROGRAMME

The Commission of the European Communities submitted to the Council on 4 November 1981 a 'Draft Action Programme of the European Communities on the Environment 1982–1986'. The document contained a draft Resolution for the Council to adopt which would approve the guidelines laid down in the annexed programme. The proposal provides for a continuation of previous programmes. The major new emphasis in the Commission's proposal is to establish clearly that the common environmental policy cannot be dissociated from measures designed to achieve the fundamental objectives of the Community. The Community therefore should seek to integrate concern for the environment into the planning and development of certain economic activities and thus promote the creation of an overall strategy making environmental policy a part of economic and social development. The original objectives and principles are reaffirmed.

The Council had an initial discussion of the proposals on 3 December 1981. There was a subsequent short discussion on 24 June 1982.

Following the opinion of the European Parliament, the Commission sent amended proposals to the Council on 21 October 1982. These were discussed by the Council on 3 December 1982 and agreed on 17 December 1982. The 17 December Resolution of the Council of the European Communities and of the Representatives of the Governments of the Member States, meeting within the Council, which took note, and approved the general approach of, the Third Environment Action Programme is set out in full in Annex VII.

2
Water

1. QUALITY OBJECTIVES

The Action Programme on the Environment of 22 November 1973 emphasized the need to establish quality objectives at the Community level. 'Quality objectives' were defined by the Council as a 'set of requirements which must be fulfilled at a given time, now or in the future, by a given environment or particular part thereof'. Quality objectives were to be set for the following uses and functions of water: drinking, swimming, farming, pisciculture and industry, beverage industry, recreation and aquatic life in general.

(a) Surface water

The Commission's first proposal (of 15 January 1974) for a Council directive on quality objectives concerned the quality required of surface water intended for the abstraction of drinking water in the Member States.

Private sector requirements for potable water within the EEC countries were assessed at an average of 100 litres per person per day. The needs of the public utilities were proportional to the populations of towns and cities and could exceed 500 litres per person per day in cities with more than a million inhabitants. Water supplies were most frequently abstracted from ground water and surface water (lakes, water courses and artificial reservoirs). Generally speaking, purification was necessary particularly in the case of river water, which was very often polluted by effluents of various origins. These effluents might contain pollutants in different concentrations, whose toxicity and harmfulness also differed.

The proposal included a detailed definition of the pollution levels which were not to be exceeded if health requirements were to be fulfilled. It laid down standard methods of treatment for transforming surface water of various categories into drinking water.

The Council adopted the draft directive on 16 June 1975. Member States were required to take all necessary measures to ensure that surface water conformed to the values laid down in the directive (see Table 2.1). They were further required to apply the directive without distinction both to national waters and to waters crossing their frontiers.

Though health requirements were a primary motivation for the directive, the ecological and social aspects were not ignored. Three categories of surface water were under normal circumstances permitted to be used for the abstraction of drinking water. These were referred to as categories A1, A2 and A3. Member States undertook to draw up a systematic plan of action, including a timetable for the improvement of surface water and especially that falling within category A3. In this context, considerable improvements were to be achieved under the national programme over the next ten years. The Commission was to carry out a thorough examination of these plans, including the timetables and would, if necessary, submit appropriate proposals to the Council. It may be seen therefore that the directive, as adopted by the Council, concerns more than health requirements. It is a tool which may be used to achieve gradual but significant improvements in the quality of surface waters, wherever these are used for the abstraction of drinking water. Because the directive applies equally to water crossing the frontiers of Member States, it must be seen as an important instrument for dealing with certain problems of trans-frontier pollution.

The directive provided that in exceptional circumstances water which did not correspond to the categories laid down in the directive might be utilized for the abstraction of drinking water on the condition that suitable processes—including blending—were used to bring the quality characteristics of the water up to the level of the quality standards for drinking water. The Commission was to be notified of the grounds for such exceptions, as soon as possible, in the case of existing installations, and in advance, in the case of new installations. Notification of such exceptions was to be made on the basis of a water resources management plan within the area concerned. The Commission would examine the grounds in detail and, when necessary, would submit appropriate proposals to the Council.

The Action Programme recognized the need to establish at a Community level a common methodology for determining the quality objectives based on the sets of reference parameters, including measuring methods for determining pollution trends in the environment in question. In this first directive the frequency of sampling and the analysis of each parameter, together with the methods of measurement was to be defined by the competent national authorities pending the development of a Community policy on the matter. The Commission prepared such a Community policy, dealing with the frequency of sampling and analysis and methods of measurement and submitted a proposal for a Directive to the Council on 25 July 1978. This proposal was adopted on 7 October 1979. Details of the reference methods of measurement and minimum annual frequency of sampling and analysis are set out in Tables 2.2 and 2.3. A consolidated report or the information provided by member states is to be published by the Commission at regular intervals. See also Section 5 on Monitoring.

Table 2.1: Council Directive of 16 June 1975
Quality requirements for surface water intended for the abstraction of drinking water

No.	Parameter	Units	A1 G	A1 I	A2 G	A2 I	A3 G	A3 I
1	pH		6.5 to 8.5		5.5 to 9		5.5 to 9	
2	Coloration (after simple filtration)	mg/l Pt scale	10	20 (0)	50	100 (0)	50–	200 (0)
3	Total suspended solids	mg/l SS	25					
4	Temperature	°C	22	25 (2)	22	25 (0)	22	25 (0)
5	Conductivity	s/cm^{-1} at 20°C	1000		1000		1000	
6	Odour	(dilution factor at 25°C)	3		10		20	
7*	Nitrates	mg/l NO_3	25	50 (0)		50 (0)		50 (0)
8[1]	Fluorides	mg/l F	0.7 to 1	1.5	0.7 to 1.7		0.7 to 1.7	
9	Total extractable organic chlorine	mg/l Cl						
10*	Dissolved iron	mg/l Fe	0.1	0.3	1	2	1	
11*	Manganese	mg/l Mn	0.05		0.1		1	
12	Copper	mg/l Cu	0.02	0.05 (2)	0.05		1	
13	Zinc	mg/l Zn	0.5	3	1	5	1	5
14	Boron	mg/l B	1		1		1	
15	Beryllium	mg/l Be						
16	Cobalt	mg/l Co						
17	Nickel	mg/l Ni						
18	Vanadium	mg/l V						
19	Arsenic	mg/l As	0.01	0.05		0.05	0.05	0.1
20	Cadmium	mg/l Cd	0.001	0.005	0.001	0.005	0.001	0.005
21	Total chromium	mg/l Cr		0.05		0.05		0.05
22	Lead	mg/l Pb		0.05		0.05		0.05
23	Selenium	mg/l Se		0.01		0.01		0.01
24	Mercury	mg/l Hg	0.005	0.001	0.005	0.001	0.0005	0.001
25	Barium	mg/l Ba		0.1		1		1
26	Cyanide	mg/l Cn		0.05		0.05		0.05
27	Sulphates	mg/l SO_4	150	250	150	250 (0)	150	250 (0)
28	Chlorides	mg/l Cl	200		200		200	
29	Surfactants (reacting with methyl blue)	mg/l (laurylsulphate)	0.2		0.2		0.5	
30*[2]	Phosphates	mg/l P_2O_5	0.4		0.7		0.7	

	Parameter		A1 G	A1 I	A2 G	A2 I	A3 G	A3 I
31	Phenols (phenol index) paranitraniline 4 aminoantipyrine	mg/l C_6H_5OH		0.001	0.001	0.005	0.01	0.1
32	Dissolved or emulsified hydrocarbons (after extraction by petroleum ether)	mg/l		0.05		0.2	0.5	1
33	Polycyclic aromatic hydrocarbons	mg/l		0.0002		0.0002		0.001
34	Total pesticides (parathion, BHC, dieldrin)	mg/l		0.001		0.0025		0.005
35*	Chemical oxygen demand (COD)	mg/l O_2					30	
36*	Dissolved oxygen saturation rate		>70		>50		>30	
37*	Biochemical oxygen demand (BOD) (at 20°C without nitrification)	mg/l O	<3		<5		<7	
38	Nitrogen by Kjeldahl method (except NO_3)	mg/l N	1		2		3	
39	Ammonia	mg/l NH_4	0.05		1	1.5	2	3 (0)
40	Substances extractable with chloroform	mg/l SEC	0.1		0.2		0.5	
41	Total organic carbon	mg/l C						
42	Residual organic carbon after flocculation and membrane filtration (5) (TOC)	mg/l C						
43	Total coliforms 37 C	/100 ml	50		5000	5000	5000	
44	Faecal coliforms	/100 ml	20		2000		2000	
45	Faecal streptococci	/100 ml	20		1000		1000	
46	Salmonella		Not present in 1000 ml		Not present in 1000 ml			

I = mandatory; G = guide; O = exceptional climatic or geographical conditions; * = see Article 8 (d)

[1] The value given are upper limits set in relation to the mean annual temperature (high and low)
[2] This parameter has been included to satisfy the ecological requirements of certain types of environment.

Table 2.2: Council directive of 9 October 1979
Measurement methods and frequency of sampling and analysis of surface water intended for the abstraction of drinking water
Reference method of measuring the I and/or G values of the parameters in Council Directive 75/440/EEC

(A)	(B) Parameter		(C) Limit of detection	(D) Precision ±	(E) Accuracy ±	(F) Reference method of measurement	(G) Materials recommended for the container
1	pH	pH unit	—	0.1	0.2	Electrometry Measured *in situ* at the time of sampling without prior treatment of the sample	
2	Coloration (after simple filtration)	mg Pt/l	5	10%	20%	Filtering through a glass fibre membrane Photometric method using the platinum-cobalt scale	
3	Total suspended solids	mg/l	—	5%	10%	Filtering through a 0.45 μm filter membrane, drying at 105°C and weighing Centrifuging (for at least 5 mins with mean acceleration of 2,800 to 3,200 g), drying at 105°C and weighing	
4	Temperature	°C	—	0.5	1	Thermometry Measured *in situ* at the time of sampling without prior treatment of the sample	
5	Conductivity at 20°C	μs/cm	—	5%	10%	Electrometry	
6	Odour	Dilution factor at 25°C	—	—	—	By successive dilutions	Glass
7	Nitrates	mg/l NO₃	2	10%	20%	Molecular absorption spectrophotometry	
8	Fluorides	mg/l F	0.05	10%	20%	Molecular absorption spectrophotometry after distillation if necessary Ion selective electrodes	

Table 2.2 continued

(A)	(B) Parameter		(C) Limit of detection	(D) Precision ±	(E) Accuracy ±	(F) Reference method of measurement	(G) Materials recommended for the container
9	Total extractable organic chlorine	mg/l Cl					
10	Dissolved iron	mg/l Fe	0.02	10%	20%	Atomic absorption spectrophotometry after filtering through a filter membrane (0.45 μm) Molecular absorption spectrophotometry after filtering through a 0.45 μm filter membrane	
11	Manganese	mg/l Mn	0.01 [2] 0.02 [3]	10%	20%	Atomic absorption spectrophotometry Atomic absorption spectrophotometry Molecular absorption spectrophotometry	
12	Copper [10]	mg/l Cu	0.005 0.02 [4]	10%	20%	Atomic absorption spectrophotometry Polarography Atomic absorption spectrophotometry Molecular absorption spectrophotometry Polarography	
13	Zinc [10]	mg/l Zn	0.01 [2] 0.02	10%	20%	Atomic absorption spectrophotometry Atomic absorption spectrophotometry Molecular absorption spectrophotometry	
14	Boron [10]	mg/l B	0.1	10%	20%	Molecular absorption spectrophotometry Atomic absorption spectrophotometry	Materials not containing boron in any significant quantities

Table 2.2 continued

(A)	(B) Parameter		(C) Limit of detection	(D) Precision ±	(E) Accuracy ±	(F) Reference method of measurement	(G) Materials recommended for the container
15	Beryllium	mg/l Be					
16	Cobalt	mg/l Co					
17	Nickel	mg/l Ni					
18	Vanadium	mg/l V					
19	Arsenic [10]	mg/l As	0.002 [2]	20%	20%	Atomic absorption spectrophotometry	
			0.01 [5]			Atomic absorption spectrophotometry Molecular absorption spectrophotometry	
20	Cadmium [10]	mg/l Cd	0.0002 0.001 [5]	30%	30%	Atomic absorption spectrophotometry Polarography	
21	Total chromium [10]	mg/l Cr	0.01	20%	30%	Atomic absorption spectrophotometry Molecular absorption spectrophotometry	
22	Lead [10]	mg/l Pb	0.01	20%	30%	Atomic absorption spectrophotometry Polarography	
23	Selenium [10]	mg/l Se	0.005			Atomic absorption spectrophotometry	
24	Mercury [10]	mg/l Hg	0.0001 0.0002 [5]	30%	30%	Flameless atomic absorption spectrophotometry (cold vaporization)	
25	Barium [10]	mg/l Ba	0.02	15%	30%	Atomic absorption spectrophotometry	
26	Cyanide	mg/l CN	0.01	20%	30%	Molecular absorption spectrophotometry	

Table 2.2 continued

(A)	(B) Parameter		(C) Limit of detection	(D) Precision ±	(E) Accuracy ±	(F) Reference method of measurement	(G) Materials recommended for the container
27	Sulphates	mg/l SO$_4$	10	10%	10%	Gravimetric analysis EDTA compleximetry Molecular absorption spectrophotometry	
28	Chlorides	mg/l Cl	10	10%	10%	Titration (Mohr's method) Molecular absorption spectrophotometry	
29	Surfactants (reacting with methylene blue)	mg/l (Lauryl Sulphate)	0.05	20%		Molecular absorption spectrophotometry	
30	Phosphates	mg/l P$_2$O$_5$	0.02	10%	20%	Molecular absorption spectrophotometry	
31	Phenols (phenol index)	mg/l C$_6$H$_5$OH	0.0005 0.001 [6]	0.0005 30%	0.0005 50%	Molecular absorption spectrophotometry 4 aminoantipyrine method Paranitraniline method	Glass
32	Dissolved or emulsified hydrocarbons	mg/l	0.01 0.04 [3]	20%	30%	Infra-red spectrometry after extraction by carbon tetrachloride Gravimetry after extraction by petroleum ether	Glass
33	Polycyclic aromatic hydrocarbons [10]	mg/l	0.00004	50%	50%	Measurement of fluorescence in the UV after thin layer chromatography Comparative measurement in relation to a mixture of six control substances with the same concentration [8]	Glass or aluminium

Table 2.2 continued

(A)	(B) Parameter	(C) Limit of detection	(D) Precision ±	(E) Accuracy ±	(F) Reference method of measurement	(G) Materials recommended for the container
34	Total pesticides (parathion, hexachloro-cyclohexane, dieldrin)[10] — mg/l	0.0001	50%	50%	Gas or liquid chromatography after extraction by suitable solvents and purification. Identification of the constituents of the mixture. Quantitative analysis[9]	Glass
35	Chemical oxygen demand (COD) — mg/l O_2	15	20%	20%	Potassium dichromate method	
36	Dissolved oxygen saturation rate — %	5	10%	10%	Winkler's method. Electrochemical method	Glass
37	Biochemical oxygen demand (BOD₅) at 20°C without nitrification — mg/l O_2	2	1.5	2	Determination of dissolved oxygen before and after five-day incubation at 20°C ± 1°C, in complete darkness. Addition of a nitrification inhibitor	
38	Nitrogen by Kjeldahl method (except in NO_2 and NO_3) — mg/l N	0.3	0.5	0.5	Mineralization, distillation by Kjeldahl method and ammonium determination by means of molecular absorption spectrophotometry or titration	
39	Ammonium — mg/l NH_4	0.01 [2] 0.1 [3]	0.03 [2] 10% [3]	0.03 [2] 20% [3]	Molecular absorption spectrophotometry	

Table 2.2 continued

(A)	(B) Parameter		(C) Limit of detection	(D) Precision ±	(E) Accuracy ±	(F) Reference method of measurement	(G) Materials recommended for the container
40	Substances extractable with chloroform	mg/l	[11]	—	—	Extraction at neutral pH value by purified chloroform, evaporation in vacuo at room temperature, weighing of residue	
41	Total organic carbon	mg/l C					
42	Residual organic carbon after flocculation and membrane filtration (5 μm)	mg/l C					
43	Total coliforms	/100 ml	5 [2] 500 [7]			Culture at 37°C on an appropriate specific solid medium (such as Tergitol lactose agar, Endo agar, 0.4% Teepol broth) with filtration [2] or without filtration [7] and colony count. Samples must be diluted or, where appropriate, concentrated in such a way as to contain between 10 and 100 colonies. If necessary, identification by gasification.	Sterilized glass
			5 [2] 500 [7]			Method of dilution with fermentation in liquid substrates in at least three tubes in three dilutions. Sub-culturing of the positive tubes on a confirmation medium. Count according to MPN (most probable number). Incubation temperature: 37°C ± 1°C.	

Table 2.2 continued

(A)	(B) Parameter	(C) Limit of detection	(D) Precision ±	(E) Accuracy ±	(F) Reference method of measurement	(G) Materials recommended for the container
44	Faecal coliforms /100 ml	2 [2] 200 [7]			Culture at 44°C on an appropriate specific solid medium (such as Tergitol lactose agar, Endo agar, 0.4% Teepol broth) with filtration [2] or without filtration [7] and colony count. Samples must be diluted or, where appropriate, concentrated in such a way as to contain between 10 and 100 colonies. If necessary, identification by gasification.	Sterilized glass
		2 [2] 200 [7]			Method of dilution with fermentation in liquid substrates in at least three tubes in three dilutions. Subculturing of the positive tubes on a confirmation medium. Count according to MPN (most probable number). Incubation temperature 44°C ± 0.5°C.	
45	Faecal streptococci /100 ml	2 [2] 200 [7]			Culture at 37°C on an appropriate solid medium (such as sodium azide) with filtration [2] or without filtration [7] and colony count. Samples must be diluted or, where appropriate, concentrated in such a way as to contain between 10 and 100 colonies.	Sterilized glass
		2 [2] 200 [7]			Method of dilution in sodium azide broth in at least three tubes with three dilutions. Count according to MPN (most probable number).	

Table 2.2 continued

(A)	(B)	(C)	(D)	(E)	(F)	(G)
	Parameter	Limit of detection	Precision ±	Accuracy ±	Reference method of measurement	Materials recommended for the container
46	Salmonella [12]	1/5000 ml 1/1000 ml			Concentration by filtration (on membrane or appropriate filter). Inoculation into pre-enrichment medium. Enrichment and transfer into isolating gels—Identification.	Sterilized glass

[1] Surface water samples taken at the abstraction point are analysed and measured after sieving (wire mesh sieve) to remove any floating debris such as wood or plastic.

[2] For waters of Category A1, G value.

[3] For waters of Categories A2 and A3.

[4] For waters of Category A3.

[5] For waters of Categories A1, A2 and A3, I value.

[6] For waters of Categories A2, I value and A3.

[7] For waters of Categories A2 and A3, G value.

[8] Mixture of six standard substances all of the same concentration to be taken into consideration: fluoranthene: 3, 4-benzofluoranthene: 11, 12-benzofluoranthene; 3, 4-benzopyrene: I. 12-benzoperylene: indano 1, 2, 3-cd pyrene.

[9] Mixture of three substances all of the same concentration to be taken into consideration: parathion, hexachlorocyclohexane, dieldrin.

[10] If the samples contain so much suspended matter as to require special preliminary treatment, the accuracy values shown in column E in this Annex may as an exception be exceeded and will be regarded as a target. These samples must be treated so as to ensure that the analysis covers the largest quantity of substances to be measured.

[11] As this method is not in current use in all the Member States, it is not certain that the limit of detection required for checking the values in Directive 75/440/EEC can be attained.

[12] Absence in 5000 ml (A1, G) and absence in 1000 ml (A2, G).

Table 2.3: Council directive of 9 October 1979
Measurement methods and frequency of sampling and analysis of surface water intended for the abstraction of drinking water

Minimum annual frequency of sampling and analysis for each parameter in Directive 75/440/EEC

Population served	A1 (*)			A2 (*)			A3 (*)		
	I (**)	II (**)	III (***)	I (**)	II (**)	III (**)	I (**)	II (**)	III (**)
≤10,000	(***)	(***)	(***)	(***)	(***)	(***)	2	1	(***)(1)
>10,000 to ≤30,000	1	1	(***)	2	1	(***)	3	1	1
>30,000 to ≤100,000	2	1	(***)	4	2	1	6	2	1
>100,000	3	2	(***)	8	4	1	12	4	1

(*)   Quality of surface waters. Annex II Directive 75/440/EEC.
(**) Classification of parameters according to frequency.
(***) Frequency to be determined by the competent national authorities.
(1)   Assuming that such surface water is intended for the abstraction of drinking water, the Member States are recommended to carry out at least annual sampling of this category of water (A3, III, 10 000).

CATEGORIES

I		II		III	
	Parameter		*Parameter*		*Parameter*
1	Ph	10	Dissolved iron	8	Fluorides
2	Coloration	11	Manganese	14	Boron
3	Total suspended solids	12	Copper	19	Arsenic
4	Temperature	13	Zinc	20	Cadmium
5	Conductivity	27	Sulphates	21	Total chromium
6	Odour	29	Surfactants	22	Lead
7	Nitrates	31	Phenols	23	Selenium
28	Chlorides	38	Nitrogen by Kjeldahl method	24	Mercury
30	Phosphates	43	Total coliforms	25	Barium
35	Chemical oxygen demand (COD)	44	Faecal coliforms	26	Cyanide
36	Dissolved oxygen saturation rate			32	Dissolved or emulsified hydrocarbons
37	Biochemical oxygen demand (BOD$_5$)			33	Polycyclic aromatic hydrocarbons
39	Ammonium			34	Total pesticides
				40	Substance extractable with chloroform
				45	Faecal streptococci
				46	Salmonella

(b) Bathing water

In February 1975 the Commission sent to the Council a second proposed directive dealing with water quality objectives. This directive related to the quality of bathing water. The directive was adopted by the Council on 8 December 1975.

In sending the proposed directive to the Council, the Commission recognized that for many years public authorities had been concerned about the part played by bathing water, particularly when polluted by sewage water, in the transmission of infectious diseases. This was not merely a matter of national concern; it was also of Community interest. Water pollution, whether of the sea or rivers, frequently had international implications. It might affect the interests of more than one Member State either because the pollution itself moved across national frontiers or because people from several Member States, especially tourists, might suffer from the effects of such pollution in the localities which they visited or frequented.

The Commission's aim in sending the directive was not purely of satisfying public health requirements. There was a growing body of public opinion in the Member States which believed that the quality of bathing water should satisfy other criteria besides those of public health, such as amenity, aesthetic attractiveness and the improvement of the quality of the environment in general. The aim of the directive was to ensure that Member States established, in accordance with certain procedures, a set of numerical values which corresponded to parameters in the directive laying down the minimum quality required of bathing water. The parameters in question were both micro-biological and physico-chemical. Certain other substances regarded as indications of pollution were also covered.

Member States agreed to take all necessary measures to ensure that, within 10 years following the notification of the directive, the quality of bathing water would conform to the limit values set in the directive (see Table 2.4). A sampling procedure was established and the minimum frequency of sampling laid down. Member States agreed to submit a comprehensive report to the Commission on their bathing water and on its most significant characteristics. With the consent of the Member State concerned, the Commission would publish the information obtained. The first such report was published by the Commission in 1982.

(c) Water for freshwater fish

On 26 July 1976 the Commission forwarded to the Council a proposal for a Council Directive on the quality requirements for water capable of supporting freshwater fish. The Commission noted that degradation in the quality of water due to the discharge of pollutants had adverse effects on certain fish populations, particularly the reduction in the overall number of certain species or even, in some cases, the disappearance of some of these species.

The quality objectives contained in this Directive aim at allowing fish belonging to indigenous species presenting a natural diversity or fish belonging to species whose presence is considered desirable for water management to live in favourable conditions. When laying down parameters, and numerical values for determining water quality as much attention as possible has been paid to the effects of each parameter not only for the survival of fish at different stages in their life cycle but also for their growth, reproduction and performance and for other components of the aquatic ecosystem which may provide them with shelter or food.

The Council adopted this Directive at its meeting on 30 May 1978. Under the Directive Member States will themselves designate the waters to which the directive applies and will fix values for the parameters listed and carry out sampling in accordance with the requirements set out (see Table 2.5). Member states will also establish programmes to reduce pollution and ensure that the designated waters conform within five years after designation with the values set by Member States and the requirements of the directive. Five years after designation Member States are to submit to the Commission a detailed report on designated waters which the Commission will publish. The first such report was published by the Commission in 1979.

(d) Water for shellfish

On 3 November 1976, the Commission sent to the Council a proposal for a Council Directive relating to the quality requirements for waters favourable to shellfish growth. The purpose of the directive was to encourage the increase of the shellfish population under suitable conditions. In fixing the parameters and numerical values which characterize the quality of the water, due account was taken as far as possible of the effects of each parameter not only on the survival of this population at the different stages of its life cycle but also on its growth and reproduction, and on other constituent elements of the aquatic ecosystem which may supply it with food.

These quality objectives were not incompatible with national or Community rules concerning consumer health protection. This proposal, however, dealt with the problems of human consumption of shellfish solely as regards the changes in the flavour of the shellfish meat brought about by certain substances.

The directive was adopted on 30 October 1979. Under the directive Member States will designate coastal and brackish waters needing protection or improvement in order to support shellfish life and growth and thus contribute to the high quality of shellfish products directly edible by man. Values for the parameters for the designated waters will be set by Member States in accordance with the directive (see Table 2.6). Member States will also establish programmes to reduce pollution and to ensure designated waters conform within six years of designation to the values set and to the requirements of the directive. Sampling will be carried out by Member States in accordance with the minimum frequency set in the directive. The

Table 2.4: Council Directive of 8 December 1975 Quality requirements for bathing water

	Parameters		G	I	Minimum sampling frequency	Method of analysis and inspection
	MICROBIOLOGICAL					
1	Total coliforms	/100 ml	500	10000	Fort-nightly (1)	Fermentation in multiple tubes subculturing of the positive tubes on a confirmation medium.
2	Faecal coliforms	/100 ml	100	2000	Fort-nightly (1)	Count according to MPN (most probable number) or membrane filtration and culture on an appropriate medium such as Tergitol lactose agar, endo agar. 0.4% Teepol broth, subculturing and identification of the suspect colonies. In the case of 1 and 2, the incubation temperature is variable according to whether total or faecal coliforms are being investigated.
3	Faecal streptococci	/100 ml	100	—	(2)	Litsky method. Count according to MPN (most probable number) or filtration on membrane. Culture on an appropriate membrane.
4	Salmonella	/1 litre	—	0	(2)	Concentration by membrane filtration. Inoculation on a standard membrane. Enrichment—subculturing on isolating—agar identification.

Table 2.4 continued

	Parameters		G	I	Minimum sampling frequency	Method of analysis and inspection
5	Entero viruses	PFU/10 litres	—	0	(2)	Concentrating by filtration, flocculation or centrifuging and confirmation.
	PHYSICO-CHEMICAL:					
6	pH		—	6 to 9 (0)	(2)	Electrometry with calibration at pH 7 and 9.
7	Colour		—	No abnormal change in colour (0)	Fortnightly (1)(2)	Visual inspection or photometry with standards on the Pt. Co scale.
8	Mineral oils	mg/litre	—	No film visible on the surface of the water and no odour —	Fortnightly (1)	Visual and olfactory inspection or extraction using an adequate volume and weighing the dry residue.
9	Surface-active substances reacting with methylene blue	mg/litre (lauryl-sulfate)	<0.3	No lasting foam	(2)	Visual inspection or absorption spectrophotometry with methylene blue.
10	Phenols (phenol indices)	mg/litre C_6H_5OH	≤0.005	No specific odour ≤0.05	Fortnightly (1)(2)	Verification of the absence of specific odour due to phenol or absorption spectrophotometry 4-aminoantipyrine (4 AAP) method.
11	Transparency	m	2	1 (0)	Fortnightly (1)	Secchi's disc.

Table 2.4 continued

	Parameters		G	I	Minimum sampling frequency	Method of analysis and inspection
12	Dissolved oxygen	% saturation O_2	80 to 190	—	(2)	Winkler's method or electrometric method (oxygen meter).
13	Tarry residues and floating materials such as wood, plastic articles, bottles, containers of glass, plastic, rubber or any other substance. Waste or splinters.		Absence		Fortnightly (1)	Visual inspection
14	Ammonia	mg/litre NH_4			(3)	Absorption spectrophotometry. Nessler's method, or indophenol blue method.
15	Nitrogen Kjeldahl	mg/litre N			(3)	Kjeldahl method
16	Pesticides (parathion, HCH, dieldrin)	mg/litre			(2)	Extraction with appropriate solvents and chromatographic determination.
17	Heavy metals such as – arsenic – cadmium – chrom VI – lead – mercury	mg/litre As Cd CrVI Pb Hg			(2)	Atomic absorption possible preceded by extraction

Table 2.4 continued

	Parameters		G	I	Minimum sampling frequency	Method of analysis and inspection
18	Cyanides	mg/litre CN			(2)	Absorption spectrophotometry using a specific reagent
19	Nitrates and phosphates	mg/litre Nl₁ PO₄			(2)	Absorption spectrophotometry using a specific reagent.

G guide; I mandatory; (0) Provision exists for exceeding the limits in the event of exceptional geographical or meteorological conditions; (1) When a sampling taken in previous years produced results which are appreciably better than those in this Annex and when no new factor likely to lower the quality of the water has appeared, the competent authorities may reduce the sampling frequency by a factor of 2; (2) Concentration to be checked by the competent authorities when an inspection in the bathing area shows that the substance may be present or that the quality of the water has deteriorated; (3) These parameters must be checked by the competent authorities when there is a tendency towards the eutrophication of the water.

Table 2.5: Council Directive of 30 May 1978
Quality requirements for freshwater needing protection or improvement in order to support fish life

	Parameter	Salmonid waters		Cyprinid waters	
		G	I	G	I

1	Temperature °C	1° – Temperature measured downstream of a point of thermal discharge (at the end of the mixing zone) must not exceed the unaffected temperature by more than:
		       1.5°C 3°C
		Derogations limited in geographical scope may be decided by Member States in particular conditions if the competent authority can prove that there are no harmful consequences for the balanced development of the fish population
		2° – Thermal discharge must not cause the temperature downstream of the point of thermal discharge (at the edge of the mixing zone) to exceed the following:
		    21.5 (0) 28 (0)
		    10 (0) 10 (0)
		The 10° temperature limit applies only to breeding periods of species which need cold water for reproduction and only to waters which may contain such species.
		Temperature limits may, however, be exceeded for 2% of the time.			

2	Dissolved oxygen mg/l O_2	50% ⩾ 9 100% ⩾ 7	50% ⩾ 9 When the oxygen concentration falls below 6 mg/l, Member States shall implement the provisions of Article 7(3). The competent authority must prove that this situation will have no harmful consequences for the balanced development of the fish population.	50% ⩾ 8 100% ⩾ 5	50% ⩾ 7 When the oxygen concentration falls below 4 mg/l, Member States shall implement the provisions of Article 7(3). The competent authority must prove that this situation will have no harmful consequences for the balanced development of the fish population.
3	pH	6–9 (0) x) x) Artificial pH variations with respect to the unaffected values shall not exceed +0.5 of a pH unit within the limits falling between 6.0 and 9.0 provided that these variations do not increase the harmfulness of other substances present in water.		6–9 (0) x)	
4	Suspended solids mg/l	⩽25 (0)		⩽25 (0)	

Methods of analysis or inspection	Minimum sampling and measuring frequency	OBSERVATIONS
Thermometry	Weekly, both upstream and downstream of the point of thermal discharge	Over-sudden variations in temperature shall be avoided
– Winkler's method or – specific electrodes (electro-chemical method)	Monthly, minimum one sample representative of low oxygen conditions of the day of sampling. However, where major daily variation are suspected, a minimum of two samples in one day shall be taken	
Electrometry calibration by means of two solutions with known pH values, preferably on either side of, and close to, the pH being measured	Monthly	
Filtration through a 0.45 µm filtering membrane, or centrifugation (5 minutes minimum, average acceleration of 2800–3200 g) drying at 105°C and weighing	The values shown are average concentration and do not apply to suspended solids with harmful chemical properties. Floods are liable to cause particularly high concentrations	

Table 2.5 continued

	Parameter	Salmonid waters		Cyprinid waters	
		G	I	G	I
5	BOD$_5$ mg/l O$_2$	≤3		≤6	
6	Total phosphorus mg/l P				
7	Nitrites mg/l NO$_2$	≤0.01		≤0.03	
8	Phenolic compounds mg/l C$_6$H$_5$OH	(*)		(*)	
9	Petroleum hydro-carbons	(*)		(*)	

For row 8: (*) Phenolic compounds must not be present in such concentrations that they adversely affect fish flavour

For row 9: (*) Petroleum products must not be present in water in such quantities that they:
form a visible film on the surface of the water or form coatings on the beds of water-courses and lakes,
– impart a detectable 'hydrocarbon' taste to fish,
– produce harmful effects in fish

Methods of analysis or inspection	Minimum sampling and measuring frequency	OBSERVATIONS
Determination of O_2 by the Winkler method before and after 5-day incubation in complete darkness at $20°\pm1°C$. (nitrification should not be inhibited)		
Molecular absoption spectrophotometry		In the case of lakes of average depth between 18 and 300 metres the following formula could be applied $$L \leqslant 10 \, \frac{\bar{Z}}{Tw} \, (1 + \sqrt{Tw})$$ L = loading expressed in mg/l per square metre lake in one year $\bar{Z}$ = mean depth of lake in metres Tw = theoretical renewal time of lake water in years. In other cases limit values of 0.2 mg/l for salmonid and of 0.4 mg/l for cyprinid water, expressed as PO_4, may be regarded as indicative in order to reduce eutrophication
Molecular absorption spectrophotometry		
By taste		An examination by taste shall be made only where the presence of Phenolic compounds is presumed
Visual By taste	Monthly	A visual examination shall be made regularly once a month, with an examination by taste only where the presence of hydrocarbon is presumed

Table 2.5 continued

	Parameter	Salmonid waters		Cyprinid waters	
		G	I	G	I
10	Non-ionized ammonia mg/l NH_3	≤0.005	≤0.025	≤0.005	≤0.025
		In order to diminish the risk of toxicity due to non-ionized ammonia, of oxygen consumption due to nitrification and eutrophication, the concentrations of total ammonium should not exceed the following:			
11	Total ammonium mg/l NH_4	≤0.04	≤1 (*)	≤0.2	≤1 (*)
		(*) In particular geographical or climatic conditions and particularly in cases of low water temperature and of reduced nitrification or where the competent authority can prove that there are no harmful consequences for the balanced development of the fish population, Member States may fix values high than 1 mg/l.			
12	Total residual chlorine mg/l HOCl	≤0.005		≤0.005	
13	Total zinc mg/l Zn	≤0.3		≤1.0	
14	Dissolved copper mg/l Cu	≤0.04		≤0.04	

Methods of analysis or inspection	Minimum sampling and measuring frequency	OBSERVATIONS
Molecular absorption spectrophotometry using indophenol blue or Nessler's method associated with pH and temperature determination	Monthly	
DPD-method (diethyl-p-phenylene-diamene)	Monthly	The I-values correspond to pH = 6. Higher concentrations of total chlorine can be accepted if the pH is higher
Atomic absorption spectrometry	Monthly	The I-values correspond to a water hardness of 100 mg/l CaCO₃. For hardness levels between 10 and 500 mg/l corresponding limit values can be found in Annex II
Atomic absorption spectrometry		The G-values correspond to a water hardness of 100 mg/l CaCO₃. For hardness levels between 10 and 500 mg/l corresponding limit values can be found in Annex II

General observations:

It should be noted that the parametric values listed in this Annex assume that the other parameters, whether mentioned in this Annex or not, are favourable. This implies, in particular, that the concentrations of other harmful substances are very low.

Where two or more harmful substances are present in mixture, joint effects (additive, synergic or antagonistic effects) may be significant.

G   guide

I   mandatory

(0)   Derogations are possible in accordance with Article II.

Table 2.6: Council Directive of 30 October 1979
Quality of shellfish waters

	Parameters	G	I	Reference methods of analysis	Minimum sampling and measuring frequency
1	pH pH unit		7–9	Electrometry Measured in situ at the time of sampling	Quarterly
2	Temperature °C	A discharge affecting shellfish waters must not cause the temperature of the waters to exceed by more than 2°C the temperature of waters not so affected		Thermometry Measured in situ at the time of sampling	Quarterly
3	Coloration (after filtration) mg Pt/l		A discharge affecting shellfish waters must not cause the colour of the waters after filtration to deviate by more than 10 mg Pt/l from the colour of waters not so affected	Filter through a 0.45 μm membrane Photometric method, using the platinum/cobalt scale	Quarterly
4	Suspended solids mg/l		A discharge affecting shellfish waters must not cause the suspended solid content of the waters to exceed by more than 30% the content of waters not so affected	Filtration through a 0.45 μm membrane, drying at 105°C and weighing Centrifuging (for at least five minutes, with mean acceleration 2,800 to 3,200 g), drying at 105°C and weighing	Quarterly
5	Salinity °/oo	12 to 38°/oo	≤40°/oo Discharge affecting shellfish waters must not cause their salinity to exceed by more than 10% the salinity of waters not so affected	Conductimetry	Monthly

Table 2.6 continued

	Parameters	G	I	Reference methods of analysis	Minimum sampling and measuring frequency
6	Dissolved oxygen saturation %	≥80%	≥70% (average value) Should an individual measurement indicate a value lower than 70%, measurements shall be repeated An individual measurement may not indicate a value of less than 60% unless there are no harmful consequences for the development of shellfish colonies	Winkler's method Electrochemical method	Monthly, with a minimum of one sample representative of low oxygen conditions on the day of sampling. However, where major daily variations are suspected, a minimum of two samples in one day shall be taken
7	Petroleum hydrocarbons		Hydrocarbons must not be present in the shellfish water in such quantities as to: produce a visible film on the surface of the water and/or a deposit on the shellfish, have harmful effects on the shellfish	Visual examination	Quarterly
8	Organohalogenated substances	The concentration of each substance in shellfish flesh must be so limited that it contributes, in accordance with Article I, to the high quality of shellfish products	The concentration of each substance in the shellfish water or in shellfish flesh must not reach or exceed a level which has harmful effects on the shellfish and larvae	Gas chromatography after extraction with suitable solvents and purification	Half-yearly

Table 2.6 continued

	Parameters	G	I	Reference methods of analysis	Minimum sampling and measuring frequency
9	*Metals* Silver Ag Arsenic As Cadmium Cd Chromium Cr Copper Cu Mercury Hg Nickel Ni Lead Pb Zinc Zn mg/l	The concentration of each substance in shell-fish flesh must be so limited that it contributes in accordance with Article I, to the high quality of shellfish products	The concentration of each substance in the shellfish water or in the shellfish flesh must not exceed a level which gives rise to harmful effects on the shellfish and their larvae The synergic effects of these metals must be taken into consideration	Spectrometry or atomic absorption preceded, where appropriate, by concentration and/or extraction	Half-yearly
10	Faecal coliforms/100 ml	≤300 in the shellfish flesh and intervalvular liquid (¹)		Method of dilution with fermentation in liquid substrates in at least three tubes in three dilutions. Subculturing of the positive tubes on a confirmation medium. Count according to MPN (most probable number). Incubation temperature $44°C \pm 0.5°C$	Quarterly
11	Substances affecting the taste of the shellfish		Concentration lower than that liable to impair the taste of the shellfish	Examination of the shellfish by tasting where the presence of one of these substances is presumed	
12	Saxitoxin (produced by dinoflagellates)				

Abbreviations:
G = guide
I = mandatory
(¹) However, pending the adoption of a Directive on the protection of consumers of shellfish products, it is essential that this value be observed in waters in which live shellfish directly edible by man.

Commission will publish information obtained from the submission by Member States of detailed reports on designated waters six years after designation and regular intervals thereafter.

(e) Drinking water

Another directive, which was proposed by the Commission on 22 July 1975 and which sought to establish both mandatory and guideline values, is that which related to the quality of water for human consumption. The directive aimed to fix levels of toxicity and noxiousness in the quality of water supplied for human consumption with reference to the most up-to-date scientific knowledge in this field. The directive as proposed laid down the maximum admissible concentrations in respect of certain parameters and the minimum required concentrations in respect of certain other parameters. It also laid down guide level values giving the concentration in water of a given substance which should ideally not be exceeded. The directive also provided that Member States should take all necessary steps to ensure the regular monitoring of the quality of water intended for human consumption.

The directive would authorize Member States to depart from the terms of the directive in order to take account of certain specific situations. The directive was finally approved by the Council at its meeting of 19 December 1978 and adopted on 15 July 1980. As adopted the directive covers all waters intended for human consumption whether supplied for that purpose used in a food production undertaking or affecting the wholesomeness of food stuff in its finished form. Natural mineral and medicinal waters are not covered (a separate directive (80/777/EEC) covers the exploitation and marketing of natural mineral waters). Member States are to fix values for the parameters set out in the directive in accordance with the requirements detailed therein (see Table 2.7). Monitoring, sampling and analysis requirements are also set.

2. STANDARDS

It is worth noting that the directives concerning water quality objectives which have been so far adopted by the Council in some respects include the setting of 'environmental quality standards'. By environmental quality standards is meant 'standards which, with legally binding force, prescribe the levels of pollution or nuisance not to be exceeded in a given environment or part thereof'. All the water quality directives have certain parameters which are mandatory or 'imperative' in effect, i.e. they must not be exceeded if the terms of the directives are to be met. Other parameters are presented as guidelines or recommended values and Member States have undertaken to endeavour to respect them as such. Discharge or emission standards are dealt with in later sections.

Table 2.7: Council Directive of 15 July 1980
Quantity of water for human consumption

LIST OF PARAMETERS

A. ORGANOLEPTIC PARAMETERS

	Parameters	Expression of the results ([1])	Guide level (GL)	Maximum admissible concentration (MAC)	Comments
1	Colour	mg/l Pt/Co scale	1	20	
2	Turbidity	mg/l SiO$_2$ Jackson units	1 0.4	10 4	Replaced in certain circumstances by a transparency test, with a Secchi disc reading in meters: GL: 6 m MAC: 2 m
3	Odour	Dilution number	0	2 at 12°C 3 at 25°C	To be related to the taste tests.
4	Taste	Dilution number	0	2 at 12°C 3 at 25°C	To be related to the odour tests.

([1]) If, on the basis of Directive 71/354/EEC as last amended, a Member State uses in its national legislation, adopted in accordance with this Directive, units of measurement other than these indicated in this Annex, the values thus indicated must have the same degree of precision.

B. PHYSICO-CHEMICAL PARAMETERS (in relation to the water's natural structure)

	Parameters	Expression of the results	Guide level (GL)	Maximum admissible concentration (MAC)	Comments
5	Temperature	°C	12	25	
6	Hydrogen ion concentration	pH unit	6.5 ≤ pH ≤ 8.5		The water should not be aggressive. The pH values do not apply to water in closed containers. Maximum admissible value: 9.5.
7	Conductivity	μS cm^{-1} at 20°C	400		Corresponding to the mineralization of the water. Corresponding relativity values in ohms/cm: 2,500.
8	Chlorides	Cl mg/l	25		Approximate concentration above which effects might occur: 200 mg/l.
9	Sulphates	SO$_4$ mg/l	25	250	See Article 8.
10	Silica	SiO$_2$ mg/l			
11	Calcium	Ca mg/l	100		
12	Magnesium	Mg mg/l	30	50	

Table 2.7b continued

	Parameters	Expression of the results	Guide level (GL)	Maximum admissible concentration (MAC)	Comments
13	Sodium	Na mg/l	20	175 (as from 1984 and with a percentile of 90)	The values of this parameter take account of the recommendations of a WHO working party (The Hague, May 1978) on the progressive reduction of the current total daily salt intake to 6 g.
				150 (as from 1987 and with a percentile of 80)	As from 1 January 1984 the Commission will submit to the Council reports on trends in the total daily intake of salt per population.
				(these percentiles should be calculated over a reference period of three years)	In these reports the Commission will examine to what extent the 120 mg/l MAC suggested by the WHO working party is necessary to achieve a satisfactory total salt intake level, and, if appropriate, will suggest a new salt MAC value to the Council and a deadline for compliance with that value. Before 1 January 1984 the Commission will submit to the Council a report on whether the reference period of three years for calculating these percentiles is scientifically well founded.
14	Potassium	K mg/l	10	12	
15	Aluminium	Al mg/l	0.05	0.2	

Table 2.7b continued

	Parameters	Expression of the results	Guide level (GL)	Maximum admissible concentration (MAC)	Comments
16	Total hardness				
17	Dry residues	mg/l after drying at 180°C		1,500	
18	Dissolved oxygen	% O_2 saturation			Saturation value >75% except for underground water.
19	Free carbon dioxide	CO_2 mg/l			The water should not be aggressive.

C. PARAMETERS CONCERNING SUBSTANCES UNDESIRABLE IN EXCESSIVE AMOUNTS [1]

	Parameters	Expression of the results [1]	Guide level (GL)	Maximum admissible concentration (MAC)	Comments
20	Nitrates	NO_3 mg/l	25	50	
21	Nitrites	NO_2 mg/l		0.1	
22	Ammonium	NH_4 mg/l	0.05	0.5	
23	Kjeldahl Nitrogen (excluding N in NO_2 and NO_3)	N mg/l		1	
24	(K Mn O_4) Oxidizability	O_2 mg/l	2	5	Measured when heated in acid medium.
25	Total organic carbon (TOC)	C mg/l			The reason for any increase in the usual concentration must be investigated.
26	Hydrogen sulphide	S µg/l		undetectable organoleptically	
27	Substances extractable in chloroform	mg/l dry residue	0.1		

[1] Certain of these substances may even be toxic when present in very substantial quantities.

Table 2.7c continued

	Parameters	Expression of the results	Guide level (GL)	Maximum admissible concentration (MAC)	Comments
28	Dissolved or emulsified hydrocarbons (after extraction by petroleum ether); Mineral oils	μg/l		10	
29	Phenols (phenol index)	C₆H₅OH μg/l		0.5	Excluding natural phenols which do not react to chlorine.
30	Boron	B μg/l	1,000		
31	Surfactants (reacting with methylene blue)	μg/l (lauryl sulphate)		200	
32	Other organochlorine compounds not covered by parameter No 55	μg/l	1		Haloform concentrations must be as low as possible.
33	Iron	Fe μg/l	50	200	
34	Manganese	Mn μg/l	20	50	

Table 2.7c continued

	Parameters	Expression of the results	Guide level (GL)	Maximum admissible concentration (MAC)	Comments
35	Copper	Cu µg/l	100 At outlets of pumping and/or treatment works and their substations 3,000 After the water has been standing for 12 hours in the piping and at the point where the water is made available to the consumer		Above 3,000 µg/l astringent taste discoloration + corrosion may occur.
36	Zinc	Zn µg/l	100 At outlets of pumping and/or treatment works and their substations 5,000 After the water has been standing for 12 hours in the piping and at the point where the water is made available to the consumer		Above 5,000 µg/l astringent taste, opalescence and sand-like deposits may occur.

Table 2.7c continued

	Parameters	Expression of the results	Guide level (GL)	Maximum admissible concentration (MAC)	Comments
37	Phosphorus	P$_2$O$_5$ µg/l	400	5,000	
38	Fluoride	F µg/l 8–12°C 25–30°C		1,500 700	MAC varies according to average temperature in geographical area concerned.
39	Cobalt	Co µg/l			
40	Suspended solids		None		
41	Residual Chlorine	Cl µg/l			See Article 8.
42	Barium	Ba µg/l	100		
43	Silver	Ag µg/l		10	If, exceptionally, silver is used non-systematically to process the water, a MAC value of 80 µg/l may be authorized.

D. PARAMETERS CONCERNING TOXIC SUBSTANCES

	Parameters	Expression of the results	Guide level (GL)	Maximum admissible concentration (MAC)	Comments
44	Arsenic	As µg/l		50	
45	Beryllium	Be µg/l			
46	Cadmium	Cd µg/l		5	
47	Cyanides	CN µg/l		50	
48	Chromium	Cr µg/l		50	
49	Mercury	Hg µg/l		1	
50	Nickel	Ni µg/l		50	
51	Lead	Pb µg/l		50 (in running water)	Where lead pipes are present, the lead content should not exceed 50 µg/l in a sample taken after flushing. If the sample is taken either directly or after flushing and the lead content either frequently or to an appreciable extent exceeds 100 µg/l, suitable measures must be taken to reduce the exposure to lead on the part of the consumer.
52	Antimony	Sb µg/l		10	
53	Selenium	Se µg/l		10	
54	Vanadium	V µg/l			

Table 2.7d continued

	Parameters	Expression of the results	Guide level (GL)	Maximum admissible concentration (MAC)	Comments
55	Pesticides and re-lated products substances considered separately	µg/l		0.1	'Pesticides and related products' means: insecticides: persistent organochlorine compounds organophosphorous compounds carbamates herbicides fungicides PCBs and PCTs
	total			0.5	
56	Polycyclic aro-matic hydrocarbons	µg/l		0.2	reference substances: fluoranthene/benzo 3.4 fluoranthene/benzo 11.12 fluoranthene/benzo 3.4 pyrene/benzo 1.12 perylene/indeno (1, 2, 3 – cd) pyrene

E. MICROBIOLOGICAL PARAMETERS

	Parameters	Results: volume of the sample in ml	Guide level (GL)	Maximum admissible concentration (MAC)	
				Membrane filter method	Multiple tube method (MPN)
57	Total coliforms (¹)	100	—	0	MPN < 1
58	Faecal coliforms	100	—	0	MPN < 1
59	Faecal streptococci	100	—	0	MPN < 1
60	Sulphite-reducing Clostridia	20	—	—	MPN ≤ 1

Water intended for human consumption should not contain pathogenic organisms.
If it is necessary to supplement the microbiological analysis of water intended for human consumption, the samples should be examined not only for the bacteria referred to in Table E but also for pathogens including:
   salmonella,
   pathogenic staphylococci,
   faecal bacteriophages,
   entero-viruses;
nor should such water contain:
   parasites,
   algas,
   other organisms such as animalcules.
(¹) Provided a sufficient number of samples is examined (95% consistent results).

	Parameters		Results: size of sample (in ml)	Guide level (GL)	Maximum admissible concentration (MAC)	Comments
61	Total bacteria counts for water supplied for human consumption	37°C	1	10 (¹) (²)	—	On their own responsibility and where parameters 57, 58, 59 and 60 are complied with, and where the pathogen organisms given above are absent, Member States may process water for their internal use the total bacteria count of which exceeds the MAC values laid down for parameter 62.
		22°C	1	100 (¹) (²)	—	
62	Total bacteria counts for water in closed containers	37°C	1	5	20	MAC values should be measured within 12 hours of being put into closed containers with the sample water being kept at a constant temperature during that 12-hour period.
		22°C	1	20	100	

(¹) For disinfected water the corresponding values should be considerably lower at the point where it leaves the processing plant.
(²) If, during successive sampling, any of these values is consistently exceeded a check should be carried out.

F. MINIMUM REQUIRED CONCENTRATION FOR SOFTENED WATER INTENDED FOR HUMAN CONSUMPTION

	Parameters	Expression of the results	Minimum required concentration (softened water)	Comments
1	Total hardness	m/l Ca	60	Calcium or equivalent cations.
2	Hydrogen ion concentration	pH		The water should not be agressive.
3	Alkalinity	mg/l HCO_3	30	
4	Dissolved oxygen			

NB: The provisions for hardness, hydrogen ion concentration, dissolved oxygen and calcium also apply to desalinated water. If, owing to its excessive natural hardness, the water is softened in accordance with Table F before being supplied for consumption, its sodium content may, in exceptional cases, be higher than the values given in the 'Maximum admissible concentration' column. However, an effort must be made to keep the sodium content at as low a level as possible and the essential requirements for the protection of public health may not be disregarded.

TABLE OF CORRESPONDENCE BETWEEN THE VARIOUS UNITS OF WATER HARDNESS MEASUREMENT

	French degree	English degree	German degree	Milligrams of Ca	Millimoles of Ca
French degree	1	0.70	0.56	4.008	0.1
English degree	1.43	1	0.80	5.73	0.143
German degree	1.79	1.25	1	7.17	0.179
Milligrams of Ca	0.25	0.175	0.140	1	0.025
Millimoles of Ca	10	7	5.6	40.08	1

3. CONTROL OF DISCHARGES OF DANGEROUS SUBSTANCES

The Action Programme established that detailed study should be given to all the possible methods, including the setting of standards for discharges, which might be necessary to achieve and maintain quality objectives 'now and in the future'. The Programme stated that priority should be given to the control of freshwater pollution by toxic, persistent and bio-accumulable substances, such as those listed in Annex I to the Convention for the Prevention of Marine Pollution by Dumping from Ships and Aircraft, signed at Oslo on 15 February 1972 and Annex I of the Convention on the Dumping of Wastes at Sea signed at London on 13 November 1972.

(a) Discharge of dangerous substances directive

On 21 October 1974 the Commission submitted to the Council a proposal for a Council decision on the reduction of pollution caused by certain dangerous substances discharged into the aquatic environment of the Community. The Commission pointed out that three conventions concerning water pollution under discussion at that time (Convention on Marine Pollution arising from land-based sources (PARIS); European Convention for the Protection of International Watercourses against Pollution (STRASBOURG); and Convention on the Protection of the Rhine against Chemical Pollution (RHINE)), in which Member States were participating, had inconsistencies between them. The purpose of the proposal was the pursuit of a coherent approach for Member States for all three conventions and to fulfil the demand of the Environment Council of November 1973 arising from the adoption of the Action Programme.

After lengthy consideration in the Environment Group of the Council, the draft decision was first considered by Ministers on 16 October 1975. The area covered by the draft decision was the whole of the aquatic environment of the Community, that is to say inland rivers and lakes, coastal waters, and territorial seas. Within those waters Member States were asked to eliminate the pollution resulting from the discharge of 'black list' substances. The black list proposed by the Commission included mercury and cadmium; organohalogen, organophosphorous and organostannic compounds, carcinogenic substances and persistent oils.

The Commission proposed a mechanism whereby concrete steps might be taken towards fulfilling this obligation to 'eliminate' pollution from the discharge of 'black list' substances. These steps included:

1. the definition on a Community basis of certain maximum 'limit values', which were to be fixed in the light of the toxicity, permanence and bioaccumulative character of the substances under consideration taking into account the best technical means available for the elimination of such substances from a discharge and

2. the obligation of the competent authorities in Member States not to exceed these Community 'limit values' whenever they give 'consent' to a discharge. (A system of prior 'consent' to any discharge was a fundamental aspect of this decision.)

There was also included in the proposed decision a 'grey list' of substances which included various dangerous metals and metalloids, such as arsenic, lead, uranium etc., biocides and their derivatives, cyanides and fluorides etc. In respect of the 'grey list' substances, Member States would undertake to lay down individual emission standards whenever they give 'consent' to, or authorize, a discharge. That individual emission standard must itself take into account relevant water quality objectives, including of course water quality objectives established on a Community basis.

Member States would be obliged to elaborate pollution reduction programmes for 'grey list' substances and to transmit them to the Commission. The Commission would in turn undertake a regular 'comparison' of these programmes, in conjunction with the Member States, so as to ensure steady and rapid progress towards attaining the stated objective of a 'severe reduction' of pollution.

After a long and sometimes heated debate the Ministers failed to agree on this proposal. The essential problem was that the British, who favoured an approach to water pollution problems based on the setting of environmental quality objectives, were not prepared to accept a system which sought to regulate pollution by discharges of 'black list' substances purely by means of emission standards.

The Commission was given the task of working out a proposal involving a compromise between the emission standards approach favoured by the majority and the quality objective argument maintained by the UK.

The Commission's proposal was that the Council would lay down quality objectives for the substances covered by the proposal as well as emission standards. The latter would apply except where Member States could prove to the Commission on the basis of an agreed monitoring procedure that the quality objectives were being met and maintained. If quality objectives could not be met or established then the emission standards would come into force.

On 8 December 1975, the Ministers agreed to seek a solution along these lines.

The Directive on pollution caused by certain dangerous substances discharged into the aquatic environment of the Community was finally adopted by the Council on 4 May 1976.

At that meeting the delegations of Belgium, Denmark, the Federal Republic of Germany, France, Ireland, Italy, Luxembourg and the Netherlands reaffirmed the opinion already expressed by their representatives at the Council meetings on 16 October and 8 December 1975 that measures against the pollution of water caused by the discharge of substances on the 'black list' could be most effectively carried out by establishing and applying

Community limit values which the national emission standards must not exceed. They stated that they would apply the directive with this in mind.

The Council and the Commission stated that the Community intended to achieve by stages the elimination of discharges of dangerous substances in the families and the groups of substances on the 'black list', taking into account the results of examinations of each of these substances by experts and the technical means available. The Commission stated that the concept of 'the best technical means available' took into account the economic availability of those means. For details of black and grey lists see Table 2.8.

(b) Groundwater discharge directive

The directive also established certain rules relating to the discharge of dangerous substances to ground water. These rules were to be of a temporary nature pending the submission by the Commission to the Council of a proposal on groundwater. The proposal for a Council Directive on the protection of groundwater against pollution caused by certain dangerous substances was submitted to the Council by the Commission on 24 January 1978. Under this proposed directive the direct discharge to groundwater of 'black list' substances would generally be banned on account of their properties, in particular acute toxicity and persistence. Indirect discharges of 'black list' substances, direct and indirect discharges of 'grey list' substances are subject to prior authorization. As part of the authorization procedure an investigation must always be made of the hydrogeological conditions obtaining in the zone in question. The application for authorization must contain information about the place and method of disposal, the essential precautions to be taken during the operation and, when appropriate, the maximum permissible concentration and quantity of a substance in the discharge.

The directive was adopted on 17 December 1979. Its purpose is to prevent pollution of groundwater by the substances in List I and List II set out in the directive (see Table 2.9). Member States are to take the necessary steps to prevent the introduction into groundwater of the substances set out in List I (black list) and to limit the introduction of List II (grey list) substances so as to avoid pollution. Any discharges are to be subject to authorization.

(c) Further work on list I substances

The Commission, assisted as appropriate by national experts, embarked on a programme of work to ensure the rapid implementation of the Council's decision of 4 May 1976. The Commission chose the following substances from the 'black list' for priority action: mercury, cadmium, aldrin, dieldrin and endrin.

Table 2.8: Council Directive of 4 May 1976 on Pollution caused by Certain Dangerous Substances discharged into the Aquatic Environment

List I of families and groups of substances

List I contains certain individual substances which belong to the following families and groups of substances, selected mainly on the basis of their toxicity, persistence and bioaccumulation, with the exception of those which are biologically harmless or which are rapidly converted into substances which are biologically harmless:
1. organohalogen compounds and substances which may form such compounds in the aquatic environment;
2. organophosphorus compounds;
3. organotin compounds;
4. substances in respect of which it has been proved that they possess carcinogenic properties in or via the aquatic environment;
5. mercury and its compounds;
6. cadmium and its compounds;
7. persistent mineral oils and hydrocarbons of petroleum origin;
and for the purposes of implementing Articles 2, 8, 9 and 14 of this Directive:
8. persistent synthetic substances which may float, remain in suspension or sink and which may interfere with any use of the waters.

List II of families and groups of substances

List II contains:
    substances belonging to the families and groups of substances in List I for which the limit values referred to in Article 6 of the Directive have not been determined;
    certain individual substances and categories of substances belonging to the families and groups of substances listed below;
and which have a deleterious effect on the aquatic environment, which can, however, be confined to a given area and which depend on the characteristics and location of the water into which they are discharged.

Families and groups of substances referred to in the second indent

1. The following metalloids and metals and their compounds:

1. zinc	6. selenium	11. tin	16. vanadium
2. copper	7. arsenic	12. barium	17. cobalt
3. nickel	8. antimony	13. beryllium	18. thalium
4. chromium	9. molybdenum	₁14. boron	19. tellurium
5. lead	10. titanium	15. uranium	20. silver

2. Biocides and their derivatives not appearing in List I.
3. Substances which have a deleterious effect on the taste and/or smell of the products for human consumption derived from the aquatic environment, and compounds liable to give rise to such substances in water.
4. Toxic or persistent organic compounds of silicon, and substances which may give rise to such compounds in water, excluding those which are biologically harmless or are rapidly converted in water into harmless substances.
5. Inorganic compounds of phosphorus and elemental phosphorus.
6. Non-persistent mineral oils and hydrocarbons of petroleum origin.
7. Cyanides, fluorides.
8. Substances which have an adverse effect on the oxygen balance, particularly: ammonia, nitrites.

[1] Where certain substances in List II are carcinogenic, they are included in category 4 of this list.

Table 2.9: Council Directive of 17 December 1979
Protection of groundwater against pollution by certain
dangerous substances

List I of families and groups of substances

List I contains the individual substances which belong to the families and groups of substances enumerated below, with the exception of those which are considered inappropriate to List I on the basis of a low risk of toxicity, persistence and bioaccumulation.

Such substances which, with regard to toxicity, persistence and bioaccumulation, are appropriate to List II are to be classed in List II.
1. Organohalogen compounds and substances which may form such compounds in the aquatic environment
2. Organophosphorus compounds
3. Organotin compounds
4. Substances which possess carcinogenic, mutagenic or teratogenic properties in or via the aquatic environment ([1])
5. Mercury and its compounds
6. Cadmium and its compounds
7. Mineral oils and hydrocarbons
8. Cyanides.

List II of families and groups of substances

List II contains the individual substances and the categories of substances belonging to the families and groups of substances listed below which could have a harmful effect on groundwater.
1. The following metalloids and metals and their compounds:

1. Zinc	6. Selenium	11. Tin	16. Vanadium
2. Copper	7. Arsenic	12. Barium	17. Cobalt
3. Nickel	8. Antimony	13. Beryllium	18. Thallium
4. Chrome	9. Molybdenum	14. Boron	19. Tellurium
5. Lead	10. Titanium	15. Uranium	20. Silver

2. Biocides and their derivatives not appearing in List I
3. Substances which have a deleterious effect on the taste and/or odour of groundwater, and compounds liable to cause the formation of such substances in such water and to render it unfit for human consumption.
4. Toxic or persistent organic compounds of silicon, and substances which may cause the formation of such compounds in water, excluding those which are biologically harmless or are rapidly converted in water into harmless substances.
5. Inorganic compounds of phosphorus and elemental phosphorus.
6. Fluorides.
7. Ammonia and nitrites.

([1]) Where certain substances in List II are carcinogenic, mutagenic or teratogenic, they are included in category 4 of this list.

The Commission sent Member States a list of 1500 substances qualifying for inclusion in List I of the Directive on dangerous substances. It followed this up with a communication listing a smaller number of substances requiring priority action because of their toxicity, persistence and bio-accumulability. This took account of the quantities produced and used and their presence in Community water. The list includes five pesticides, the polychlorinated biphenyls (PCB's) and carcinogenic substances. A special study is being carried out on the latter to identify them and isolate those which exhibit carcinogenic properties either in or via the aquatic environment (see Table 2.10). The Commission with the collaboration of national experts is continuing to implement the inventory of substances in List I provided for in the directive. On 14 June 1982 it sent to the Council a communication on dangerous substances which might be included in List I of the directive (see Table 2.11).

At their meeting of 17 December 1982, EEC Environment Ministers in a Council Resolution took note of the Commission's communication and stated that the list of 129 substances contained in the communication would serve as a basis for further community work on the implementation of Directive 76/464/EEC (discharge of dangerous substances). They also welcomed the fact that Member States would endeavour to communicate to the Commission as soon as possible, and within three years at the latest, all readily available data concerning the list of 129 substances. When such data were being forwarded, prominence should be given to the following points:

- production, use and discharges as per branch of industry
- diffuse sources
- concentration in surface water, sediments and organisms;
- remedial measures already taken and/or envisaged and their effect on the quantities discharged

Initially, special attention would, as far as possible and where appropriate, be focussed on:

2-chloroaniline,
3-chloroaniline,
4-chloroaniline,
1-chloro-2-nitrobenzene,
1-chloro-3-nitrobenzene,
1-chloro-4-nitrobenzene,
2, 4-dichlorophenol,
2-chloroethanol,
1, 3-dichloro-2-propanol,
epichlorohydrin,
parathion (including methyl parathion).

(d) Discharges of aldrin, dieldrin and endrin

The Commission submitted to the Council on 16 May 1979 the first proposals for black list substances under the framework directive; these were two draft directives setting, respectively, limit values and quality objectives for aldrin, dieldrin and endrin. These pesticides are manufactured in only one plant in the Community and 95% of discharges to water occurs in the

Table 2.10: Council Directive 76/464/EEC
Substances selected so far and progress made

Substance	Progress made
First series:	
1. Mercury and mercury compounds	Proposal for a directive concerning the chloralkali electrolysis industry sent to the Council on 20 June 1979, adopted on 22.03.1982 (OJ No L 81 of 27.03.1982). Proposal for a directive on other industries in preparation.
2. Cadmium and cadmium compounds	Proposal for a directive sent to the Council on 17 February 1981 (OJ No C 118 of 21.5.81).
3. Aldrin 4. Dieldrin 5. Endrin	Proposal for a directive sent to the Council on 16 May 1979 (OJ No C 146 of 12.6.1979).
Second series:	
6. Chlordane 7. Heptachlor (including Heptachlorepoxide)	Communication by the Commission to the Council of 18 July 1980 (COM(80) 433 final) of which the Council took formal note on 3 December 1981.
8. DDT 9. Hexachlorocyclohexane (including all of the isomers and in particular Lindane)	The studies and discussions with the national experts are now completed. Appropriate proposals in preparation.
10. PCBs (including PCTs) 11. Hexachlorobenzene	Studies completed: discussions under way with the national experts.
Third series:	
12. Endosulfan 13. Hexachlorobutadiene 14. Pentachlorophenol 15. Trichlorophenol	Studies completed. Discussions under way with the national experts.
Fourth series:	
16. Benzene 17. Carbon tetrachloride 18. Chloroform	Studies in progress.
Carcinogens:	
19. Arsenic and mineral compounds of Arsenic 20. Benzidine 21. PAH (in particular 3,4 Benzopyrene and 3,4 Benzofluoranthene)	Studies in progress.

Table 2.11: List of substances which could belong to List I of Council Directive 76/464-EEC

1	***	309-00-2	Aldrin
2		95-85-2	2-Amino-4-chlorophenol
3		120-12-7	Anthracene
4	**	7440-38-2	Arsenic and its mineral compounds
5		2642-71-9	Azinphos-ethyl
6		86-50-0	Azinphos-methyl
7	**	71-43-2	Benzene
8	**	92-87-5	Benzidine
9		100-44-7	Benzyl chloride (Alpha-chlorotoluene)
10		98-87-3	Benzylidene chloride (Alpha, alpha-dichlorotoluene)
11		92-52-4	Biphenyl
12	***	7440-43-9	Cadmium and its compounds
13	**	56-23-5	Carbon tetrachloride
14		302-17-0	Chloral hydrate
15	***	57-74-9	Chlordane
16		79-11-8	Chloroacetic acid
18		108-42-9	3-Chloroaniline
19		106-47-8	4-Chloroaniline
20	*	108-90-7	Chlorobenzene
21		97-00-7	1-Chloro-2,4-dinitrobenzene
22		107-07-3	2-Chloroethanol
23	**	67-66-3	Chloroform
24		59-50-7	4-Chloro-3-methylphenol
25		90-13-1	1-Chloronaphthalene
26			Chloronaphthalenes (technical mixture)
27		89-63-4	4-Chloro-2-nitroaniline
28		89-21-4	1-Chloro-2-nitrobenzene
29		88-73-3	1-Chloro-3-nitrobenzene
30		121-73-3	1-Chloro-4-nitrobenzene
31		89-59-8	4-Chloro-2-nitrotoluene
32			Chloronitrotoluenes (other than 4-Chloro-2-nitrotoluene)
33		95-57-8	2-Chlorophenol
34		108-43-0	3-Chlorophenol
35		106-48-9	4-Chlorophenol
36		126-99-8	Chloroprene (2-Chlorobuta-1,3-diene)
37		107-05-1	3-Chloropropene (Allyl chloride)
38		95-49-8	2-Chlorotoluene
39		108-41-8	3-Chlorotoluene
40		106-43-4	4-Chlorotoluene
41			2-Chloro-p-toluidine
42			Chlorotoluidines (other than 2-Chloro-p-toluidine)
43		56-72-4	Coumaphos
44		108-77-0	Cyanuric chloride (2,4,6-Trichloro-1,3,5-triazine)
45		94-75-7	2,4-D (including 2,4-D-salts and 2,4-D-esters)
46	**	50-29-3	DDT (including metabolites DDD and DDE)
47		298-03-3	Demeton (including Demeton-o, Demeton-s, Demeton-s-methyl and Demeton-s-methylsulphone)
48	*	106-93-4	1,2-Dibromoethane
49			Dibutyltin dichloride
50			Dibutyltin oxide

Table 2.11 continued

51			Dibutyltin salts (other than Dibutyltin dichloride and Dibutyltin oxide)
52			Dichloroanilines
53		95-50-1	1,2-Dichlorobenzene
54		541-73-1	1,3-Dichlorobenzene
55		106-46-7	1,4-Dichlorobenzene
56			Dichlorobenzidines
57		108-60-1	Dichlorodiisopropyl ether
58	*	75-34-3	1,1-Dichloroethane
59	*	107-06-2	1,2-Dichloroethane
60	*	75-35-4	1,1-Dichloroethylene (Vinylidene chloride)
61	*	540-59-0	1,2-Dichloroethylene
62	*	75-09-2	Dichloromethane
63			Dichloronitrobenzenes
64		120-83-2	2,4-Dichlorophenol
65	*	78-87-5	1,2-Dichloropropane
66		96-23-1	1,3-Dichloropropan-2-ol
67		542-75-6	1,3-Dichloropropene
68		78-88-6	2,3-Dichloropropene
69		120-36-5	Dichlorprop
70		62-73-7	Dichlorvos
71	***	60-57-1	Dieldrin
72		109-89-7	Diethylamine
73		60-51-5	Dimethoate
74		124-40-3	Dimethylamine
75		298-04-4	Disulfoton
76	**	115-29-7	Endosulfan
77	***	72-20-8	Endrin
78		106-89-8	Epichlorohydrin
79		100-41-4	Ethylbenzene
80		122-14-5	Fenitrothion
81		55-38-9	Fenthion
82	***	76-44-8	Heptachlor (including Heptachlorepoxide)
83	**	118-74-1	Hexachlorobenzene
84	**	87-68-3	Hexachlorobutadiene
85	**	608-73-1 58-89-9	Hexachlorocyclohexane (including all isomers and Lindane)
86		67-72-1	Hexachloroethane
87		98-83-9	Isopropylbenzene
88		330-55-2	Linuron
89	*	121-75-5	Malathion
90		94-74-6	MCPA
91		93-65-2	Mecoprop
92	***	7439-97-6	Mercury and its compounds
93		10265-92-6	Methamidophos
94		7786-34-7	Mevinphos
95		1746-81-2	Monolinuron
96		91-20-3	Naphthalene
97		1113-02-6	Omethoate
98		301-12-2	Oxydemeton-methyl

Table 2.11 continued

99	**		PAH (with special reference to: 3,4-Benzopyrene and 3,4-Benzofluoranthene)
100		56-38-2 298-00-0	Parathion (including Parathion-methyl)
101	**		PCB (including PCT)
102	**	87-86-5	Pentachlorophenol
103		14816-18-3	Phoxim
104		709-98-8	Propanil
105		1698-60-8	Pyrazon
106		122-34-9	Simazine
107		93-76-5	2,4,5-T (including 2,4,5-T salts and 2,4,5-T esters)
108			Tetrabutyltin
109		95-94-3	1,2,4,5-Tetrachlorobenzene
110	*	79-34-5	1,1,2,2-Tetrachloroethane
111	*	127-18-4	Tetrachloroethylene
112		108-88-3	Toluene
113		24017-47-8	Triazophos
114		126-73-8	Tributyl phosphate
115			Tributyltin oxide
116		52-68-6	Trichlorfon
117	*		Trichlorobenzene (technical mixture)
118		120-82-1	1,2,4-Trichlorobenzene
119	*	71-55-6	1,1,1-Trichloroethane
120	*	79-00-5	1,1,2-Trichloroethane
121	*	79-01-6	Trichloroethylene
122	**	95-95-4 88-06-2	Trichlorophenols
123		76-13-1	1,1,2-Trichlorotrifluoroethane
124		1582-09-8	Trifluralin
125		900-95-8	Triphenyltin acetate (Fentin acetate)
126			Triphenyltin chloride (Fentin chloride)
127		76-87-9	Triphenyltin hydroxide (Fentin hydoxide
128		75-01-4	Vinyl chloride (Chloroethylene)
129			Xylenes (technical mixture of isomers)

*** Substances which are the subject of a proposal or a communication to the Council.
** Substances which have been or are being studied.
* Substances to be studied next.
309-00-2 CAS number (Chemical Abstract Service)

production or in the moth-proofing of wool and wool products. The proposals were therefore limited to these two industries. No agreement could be reached in Council, but the Commission made known that it would be making proposals for a simplified procedure where, as in this case, discharges took place in only one, two or three member states.

(e) Discharges of mercury for chlor-alkali plants

On 26 June 1979 the Commission submitted to the Council separate proposals for limit values and for quality objectives applicable to discharges of mercury to water by the chlor-alkali industry. The purpose of both directives was to eliminate pollution caused by such discharges. The proposals resulted in a long debate, the main problem being the treatment of new plant. Agreement was finally reached on 3 December 1981 and a single directive covering both limit values and quality objectives was adopted on 22 March 1982. The directive sets limit values to be met by the authorizations by member states except where a member state uses quality objectives as provided in the parent directive. In such a case it will do so on the basis of the requirements concerning quality objectives and monitoring now set (see details in Table 2.12). Member states may grant authorizations for new plants only if these contain a reference to the standards corresponding to the best technical means available for preventing discharges of mercury. Where for technical reasons the intended measures do not conform to the best technical means available, the member state must provide the Commission with the justification for these reasons. In an accompanying statement published in the Official Journal, the Council and the Commission state that the application of the best technical means available makes it possible to limit discharges of mercury for the site of a new industrial plant using the recycled brine process to less than 0.5 grams per tonne of installed chlorine capacity.

(f) Mercury discharges by sectors other than the chlor-alkali electrolysis industry

On 15 December 1982, the Commission sent a proposal for a Council directive on limit values and quality objectives for mercury discharges by sectors other than the chlor-alkali electrolysis industry.

The proposed Directive completes Directive 82/176/EEC of 22 March 1982—its purpose is to eliminate pollution caused by discharges containing mercury. It applies to discharges from the sectors of industry listed in Table 2.13.

All discharges containing mercury would have to comply with emission standards which meet the limit values laid down in the Directive. These limit values were determined primarily on the basis of the toxicity, persistence and bio-accumulation of mercury. They took due account of the best technical means available for effluent treatment and were determined for each sector and type of product.

A Member State might, by way of exception, draw up emission standards based on the quality objectives laid down in respect of mercury present in the areas affected by the discharges. These quality objectives for mercury were defined in Directive 82/176/EEC (see Table 2.12) and the same quality objectives also applied to discharges from the other industrial sectors covered by the proposal.

Table 2.12: Council Directive of 22 March 1982

Limit values and quality objectives for mercury discharges by the chlor-alkali industry

I. Limit values, time limits by which they must be complied with, and monitoring procedure for discharges			

1. The limit values expressed in terms of concentration which, in principle, should not be exceeded are set out in the following table.

Unit of measurement	*Monthly average limit values not to be exceeded from 1 July*		*Remarks*
	1983	*1986*	
Recycled brine and lost brine Micrograms of mercury per litre	75	50	Applicable to the total quantity of mercury present in all mercury-containing water discharged from the site of the industrial plant

In all cases, limit values expressed as maximum concentrations may not be greater than those expressed as maximum quantities divided by water requirements per tonne of installed chlorine production capacity.

2. However, because the concentration of mercury in effluents depends upon the volume of water involved, which is different for different processes and plants, the limit values expressed in terms of quantity of mercury discharged in relation to installed chlorine production capacity given in the following table must be observed in all cases.

Unit of measurement	*Monthly average limit values not to be exceeded from 1 July*		*Remarks*
	1983	*1986*	
Recycled brine Grams of mercury per tonne of installed chlorine production capacity	0.5	0.5	Applicable to the mercury present in effluent discharged from the chlorine production unit
	1.5	1.0	Applicable to the total quantity of mercury present in all mercury-containing water discharged from the site of the industrial plant
Lost brine Grams of mercury per tonne of installed chlorine production capacity	8.0	5.0	Applicable to the total quantity of mercury present in all mercury-containing water discharged from the site of the industrial plant

3. The daily average limit values are four times the corresponding monthly average limit values given in points 1 and 2.

Table 2.12 continued

4.   In order to check whether the discharges comply with the emission standards which have been fixed in accordance with the limit values laid down in this Annex, a monitoring procedure must be instituted. This procedure must provide for:
    the taking each day of a sample representative of the discharge over a period of 24 hours and the measurement of the mercury concentration of that sample, and
    the measurement of the total flow of the discharge over that period.
The quantity of mercury discharged during a month must be calculated by adding together the quantities of mercury discharged each day during that month. This total must then be divided by the installed chlorine production capacity.

II. Quality objectives

For those Member States which apply the exception provided for in Article 6 (3) of Directive 76/464/EEC, the emission standards which Member States must establish and ensure are applied, pursuant to Article 5 of that Directive, shall be fixed so that the appropriate quality objective or objectives from among those listed below is or are complied with in the area affected by discharges of mercury from the chlor-alkali electrolysis industry. The competent authority shall determine the area affected in each case and shall select from among the quality objectives listed in paragraph 1 the objective or objectives that it deems appropriate having regard to the intended use of the area affected, taking account of the fact that the purpose of this Directive is to eliminate all pollution.

1.   In order to eliminate pollution as defined in Directive 76/464/EEC, and pursuant to Article 2 of that Directive, the following quality objectives are set:

1.1. The concentration of mercury in a representative sample of fish flesh chosen as an indicator must not exceed 0.3 mg/kg wet flesh.

1.2. The total concentration of mercury in inland surface waters affected by discharges must not exceed 1 μg/l as the arithmetic mean of the results obtained over a year.

1.3. The concentration of mercury in solution in estuary waters affected by discharges must not exceed 0.5 μg/l as the arithmetic mean of the results obtained over a year.

1.4. The concentration of mercury in solution in territorial sea waters and internal coastal waters other than estuary waters affected by discharges must not exceed 0.3 μg/l as the arithmetic mean of the results obtained over a year.

1.5. The quality of the waters must be sufficient to comply with the requirements of any other Council Directive applicable to such waters as regards the presence of mercury.

2.   The concentration of mercury in sediments or in shellfish must not increase significantly with time.

3.   Where several quality objectives are applied to waters in an area, the quality of the waters must be sufficient to meet each of them.

4.   The numerical values of the quality objectives specified in 1.2, 1.3 and 1.4 may, as an exception and where this is necessary for technical reasons, be multiplied by 1.5 until 30 June 1986, provided that the Commission has been notified beforehand.

The proposal also set the dates by which existing establishments must comply with the limit values. New establishments would be required to meet emission standards making reference to the best technical means available at the time they become operational.

The Commission proposed that the sampling procedures and the reference methods of analysis used to measure the concentration of mercury which were defined in the directive on discharges of mercury from the chlor-alkali electrolysis industry should also apply to the monitoring of compliance with the emission standards laid down for discharges by other sectors.

Table 2.13: Mercury discharges by sectors other than the Chlor-Alkali Electrolysis Industry

Limit values, time limits by which they must be complied with, and monitoring procedure for discharges

1. *The limit values and the time limits for compliance for the sectors concerned are set out below:*

Sector	Limit value which must be respected by stated date:		Unit of measurement
	1 January 1985	1 January 1988	
1	2	3	4
1. Chemical industries using mercury catalysts	10	5	g/kg mercury used
	0.1	0.05	mg/l effluent
2. Manufacture of industrial catalysts which contain mercury	8	5	g/kg mercury processed
	0.1	0.05	mg/l effluent
3. Manufacture of organic and non-organic mercury compounds (excepting products in Group 2 above)	0.2	0.1	g/kg mercury processed
	0.1	0.05	mg/l effluent
4. Manufacture of mercury batteries	0.1	0.05	g/kg mercury processed
	0.1	0.05	mg/l effluent
5. Non-ferrous metal industries			
5.1 Mercury recovery plants	0.1	0.05	mg/l effluent
5.2 Extraction and refining of non-ferrous metals	0.1	0.05	mg/l effluent
6. Plants for the treatment of toxic wastes	0.1	0.05	mg/l effluent
7. Analytical laboratories using mercury reagents where the monthly rate of discharge of mercury is greater than 100 g	0.1	0.05	mg/l effluent

The limit values given in the table correspond to a monthly average concentration or to a maximum monthly load.

The amounts of mercury discharged are expressed as a function of the amount of mercury used or processed by the industrial establishment over the same period.

2. The limit values expressed as maximum concentrations may not be greater than those expressed as maximum quantities divided by the water requirements per kg of mercury

used or processed. However, because the concentration of mercury in effluents depends upon the volume of water involved—which is different for different processes and plants—the limit values expressed in terms of the quantity of mercury discharged in relation to the quantity of mercury used or processed given in the table must be observed in all cases.

3. The daily average limit values are twice the corresponding monthly average limit values given in the Table.
4. In order to check whether the discharges comply with the emission standards which have been fixed in accordance with the limit values laid down in this Table, a monitoring procedure must be instituted. This procedure must provide for:

   • the taking each day of a sample representative of the discharge over a period of 24 hours and the measurement of the mercury concentration in that sample, and
   • the measurement of the total flow of the discharges over that period.

The quantity of mercury discharged during a month must be calculated by adding together the quantities of mercury discharged each day during that month. This total must then be divided by the weight of mercury processed during that month.

Quality objectives

For those Member States which apply the exception provided for in Article 6(3) of Directive 76/464/EEC, the emission standards which Member States must establish and ensure are applied, pursuant to Article 5 of that Directive, shall be fixed so that the appropriate quality objective(s) from among those listed in Annex II to Directive 82/176/EEC is(are) complied with in the area affected by discharges of mercury emanating from the industries listed in Annex I to the present Directive. The competent authority shall determine the area affected in each case and shall select from among the quality objectives listed in Annex II.1 to Directive 82/176/EEC the objective(s) that it deems appropriate having regard to the intended use of the area affected, taking account of the fact that the purpose of this Directive is to eliminate all pollution.

The numerical values of the quality objectives specified in 1.2, 1.3 and 1.4 of Annex II to Directive 82/176/EEC may, as an exception and where this is necessary for technical reasons, be multiplied by 1.5 until 30 June 1988, provided that the Commission has been notified beforehand.

(g) Discharges of cadmium

On 17 February 1981 the Commission submitted to the Council its proposal for a directive for limit values for discharges of cadmium to water and quality objectives for cadmium in water. The purpose of the directive was to eliminate pollution caused by discharges containing cadmium. The directive would apply to all industrial discharges except for those from the manufacture of phosphoric acid from phosphate rock for which it was not practicable to set limit values. Two sets of limit values were proposed: the first which would come into operation on 1 January 1983 was based on existing good practice; the second which would apply from 1 January 1986, was based on the best technical means at present available. The quality objectives proposed were set separately for fresh water and salt water and provided for maximum permissible concentrations of cadmium, no significant increase with time in molluscs or alternatively in sediments, and protection of health of persons consuming fish from the water in question and of any other legitimate use of such water. The proposal is still under discussion in the Council (see Tables 2.14 and 2.15).

Table 2.14: Commission Proposal of 12 February 1981 concerning limit values for discharges of cadmium into the aquatic environment and quality objectives for cadmium in the aquatic environment

Limit values, time-limits and verification frequencies and procedures for discharges of cadmium

Industry	Unit of measurement	Limit values which must be respected by stated dates		Verification frequency
		1.1.1983	1.1.1986	
Zinc mining, lead and zinc refining and the non-ferrous metal industry	Milligrams of cadmium per litre of discharge	1.0 (1) 0.5 (2)	0.6 (1) 0.3 (2)	daily monthly
Manufacture of pigments	milligrams of cadmium per litre of discharge	2.0 (1) 1.0 (2)	1.0 (1) 0.5 (2)	daily monthly
	grams of cadmium discharged per kilogram of cadmium handled	1.4 (3) 0.7 (4)	0.6 (3) 0.3 (4)	daily monthly
Manufacture of stabilizers	milligrams of cadmium per litre of discharge	2.0 (1) 1.0 (2)	1.0 (1) 0.5 (2)	daily monthly
	grams of cadmium per kilogram of cadmium handled	1.6 (3) 0.8 (4)	1.0 (3) 0.5 (4)	daily monthly
Manufacture of batteries	milligrams of cadmium per litre of discharge	2.0 (1) 1.0 (2)	1.0 (1) 0.5 (2)	daily monthly
	grams of cadmium per kilogram of cadmium handled	5.0 (3) 2.5 (4)	3.0 (3) 1.5 (4)	daily monthly
Electroplating	milligrams of cadmium per litre of discharge	2.0 (1) 1.0 (2)	1.0 (1) 0.5 (2)	daily monthly
	grams of cadmium per kilogram of cadmium handled	1.0 (3) 0.5 (4)	0.6 (3) 0.3 (4)	daily monthly
Manufacture of cadmium compounds	milligrams of cadmium per litre of discharge	2.0 (1) 1.0 (2)	1.0 (1) 0.5 (2)	daily monthly
	grams of cadmium per kilogram of cadmium handled	2.0 (3) 1.0 (4)	1.0 (3) 0.5 (4)	daily monthly

Table 2.14 continued

Other industries, except the manufacture of phosphoric acid from phosphate rock	milligrams of cadmium per litre of discharge	2.0	(1)	1.0	(1)	daily
		1.0	(2)	0.5	(2)	monthly

(1) Maximum daily average concentration
(2) Maximum monthly average concentration
(3) Maximum daily load
(4) Maximum monthly load

If the discharge contains a contribution from a process or processes in which cadmium is not handled the limit value for monthly average concentration of cadmium must be calculated from the following formula:

$$C = Lv/V$$

where

C is the limit value to be applied; L is the appropriate limit value taken from the above table; v is the total monthly flow of discharge attributable to the handling of cadmium; V is the total monthly flow of the discharge in question.

In this case the limit value for daily average concentration is twice the limit value for monthly average concentration.

(h) Heptachlor and chlordane

The Council on 3 December 1981 took note of a Commission communication that no useful purpose could be served in submitting proposals for controlling discharges of heptachlor and chlordane.

(i) Reduction programme for List II substances

In 1978, the Commission organized preliminary meetings to compare national programmes for the reduction of pollution by substances on List II. The reduction programmes for List II substances have not yet been submitted in all cases and the Commission has not been able to compare the programmes to ensure uniform implementation as proposed in Article 7 of the parent directive.

4. INDUSTRIAL SECTORS

The Action Programme stated that protection of the environment required that particular attention be paid to industrial activities in which the manufacturing processes entailed the introduction of pollutants or nuisances into the environment. It was appropriate therefore:

Table 2.15: Quality objectives for cadmium in water

1.	*Fresh water*

1.1. The maximum permissible concentration of cadmium must be set in relation to the hardness of the water in question and must not exceed the following values:

Hardness of water (mg/l as $CaCO_3$)	Maximum cadmium concentration ($\mu g/l$)
<10	0.6
≥10 to <50	0.8
≥50 to <100	1.0
≥100	1.5

1.2. The cadmium content of sediments or of a characteristic mollusc must not increase *significantly with time*. The choice between these two possibilities must be made by the competent national authorities according to local circumstances.

1.3. The quality objective must be such as to protect the health of persons consuming fish taken from the water in question and to protect any other legitimate use of such water.

2.	*Salt water*

2.1. The cadmium content of salt water shall not exceed 1 $\mu g/l$. Where this quality objective cannot be achieved because of existing concentrations of cadmium the quality objective shall be that the cadmium concentration must not be more than 1 $\mu g/l$ higher than that in an adjacent area of unpolluted salt water.

2.2. The cadmium content of sediments or of a characteristic mollusc must not increase significantly with time. The choice between these two possibilities must be made by the competent national authorities according to local circumstances.

2.3. The quality objective must be such as to protect the health of persons consuming fish taken from the water in question and to protect any other legitimate use of such water.

3. It shall be deemed that the quality objectives specified in paragraphs 1.1 and 2.1 have been satisfied if during the course of any period of one year, at least ninety five percent of the relevant samples comply with the appropriate quality objective.

4. Higher cadmium concentration than those referred in paragraphs 1.1 and 2.1 shall not be taken into account when they arise as a result of floods, natural disasters or exceptional meteorological conditions.

5. The quality objectives given in paragraphs 1.1 and 2.1 apply from 1 January 1986. Less severe quality objectives apply from 1 January 1983 but in no case may these exceed twice the relevant value specified in 1.1 and 2.1.
The quality objectives specified in 1.2, 1.3, 2.2 and 2.3 apply from 1 January 1983.

6. When as a result of changes in the pattern of industry areas of water are affected by discharges from new plant, the quality objectives in paragraphs 1.2 and 2.2 shall apply two years after the start of the new discharge.

7. Where a stretch of water has more than one use which is to be protected, the quality objective must be stringent enough to protect all these uses.

● to endeavour to work out technical or other measures which could reduce, eliminate or prevent the pollutant emissions or nuisances stemming from each of the polluting industries;

● to study ways and means of implementing these measures, particularly
as regards their phasing, account being taken of existing circumstances,
the state of the art and the economic, financial and social consequences
of the measures planned.

The pulp sector of the paper and pulp industry was to be looked at as a
matter of priority, due to the potentially highly polluting nature of the
manufacturing processes used.

(a) Paper pulp industry

On 14 January 1975 the Commission submitted to the Council a Technical
Report on Pollution of Water by Pulp Manufacturing Industry in the EEC.
This report showed that pulp mill effluent could contain appreciable quantities of suspended solids, could severely deplete the oxygen content of the
receiving water-course, could contain toxic substances and could discolour
and cause foaming in the receiving water-course.

Whether or not this potential to pollute was realized, however, would
depend on:

● the type of pulp producing process employed;
● the volume and type of discharge;
● the environmental characteristics of the receiving medium;
● the extent to which Member States had established legislation controlling the discharge of waste.

At the same time as it submitted its Technical Report, the Commission
also presented to the Council a proposal for a Council directive on the
reduction of water pollution caused by paper pulp mills in the Member
States.

In preparing this directive, the Commission was guided by the general
principles defined in the Communities' 'Programme of Action on the Environment', which stressed in particular that 'the best environmental policy
consists in preventing the creation of pollution or nuisances at source, rather
than subsequently trying to counteract their effects', and also that: 'major
aspects of environmental policy in individual countries must no longer be
planned and implemented in isolation. On the basis of a common long term
concept, national programmes in these fields should be coordinated, and
national policies harmonized within the Community'. In the Commission's
view coordination and harmonization of policies in the case of the pulp
industry must initially mean the establishment of certain minimum effluent
emission limits, which were technically feasible and economically realistic
and which would represent an important first step in the protection of the
environment. In its directive the Commission therefore proposed the adoption, on a Community basis, of minimum emission standards for the pulp
industry, according to the type of manufacturing process employed.

To allow the assimilative capacity of the receiving waters, where this exists, to be nevertheless taken into account—as well as appropriate water quality criteria and local social and economic conditions—a certain measure of flexibility in applying the proposed standards was provided for. This would enable national and local authorities to work out programmes of discharge reduction, if necessary on a case by case basis.

The emission standards proposed were considered to be reasonable for mills discharging into inland waters from the environmental, as well as economic and technical view point. However, it was recognized that the assimilative capacity of tidal waters could be substantially different from those of rivers, and furthermore, that the parameters which determine the effects of effluent discharge into such waters might not be the same as in the case of rivers. It was therefore proposed that those existing mills whose discharge into tidal waters caused no appreciable damage to the environment, might be exempt from compliance with the discharge norms laid down in the directive. Such exemptions were to be temporary.

The Commission recognized that the application of the proposed discharge norms might in some cases cause undesirable economic problems and necessitate certain special aids. It referred in this connexion to a communication it had made to the Member States on this subject on 4 November 1974, regarding the Community approach to state aids in environment matters (see page 166).

The Council considered, but failed to agree on the draft directive, at its meeting of 14 June 1977. The proposal remains unadopted.

(b) Titanium dioxide industry

(1) Technical report

Another industrial sector which the Action Programme of the Environment indicated should be studied as a matter of priority was the manufacture of titanium dioxide. On 14 July 1975 the Commission presented a technical report to the Council on pollution caused by the titanium dioxide industry. The report noted that the production capacity of the titanium dioxide factories in the nine countries of the European Community was 840,000 tonnes per year (tpa). This represented 39% of world's capacity (2,175,000 tpa) and was divided up as follows:

741,000 tpa (88%) for the sulphate process
99,000 tpa (12%) for the chloride process.

The vast majority of factories manufacturing titanium dioxide dumped their waste at sea or in estuaries, relying on the buffer effect of the sea to neutralize the acid part of the waste and on the capacity of the oxygen present to convert the ferrous-sulphate to ferric-sulphate, the other waste (various oxides of heavy metals) sinking naturally to the seabed.

Factories discharging into the Channel or the North Sea accounted for 727,000 tpa (87%) and factories discharging into the Mediterranean accounted for 50,000 tpa (6%). Two factories (63,000 tpa or 7%) treated their waste on land.

The report noted that a whole series of ecological pollution monitoring campaigns had been carried out in the actual discharge areas; these campaigns had been sponsored by the national authorities or sometimes by the manufacturers themselves. In the Commission's view, it was clear from an analysis of the results of this monitoring that waste from the TiO_2 industry was potentially or actually harmful. The adverse effects on the marine environment were due above all to acidity, the presence of ferrous sulphate and probably other metals (heavy metals).

The effects in question could take the following various forms, depending on the method and place of dumping:

(a) reduced oxygenation and pH of the water and increased concentration of Fe and heavy metals:

(b) 1. temporary shortage of the zooplankton biomass and inducement of effects leading to a deterioration of the morphological structure of its components;
     2. repulsion and loss of some species of fish;
     3. reduction of the biomass, production and specific diversity of benthic and/or nectobenthic biocenoses in the discharge area. In more severe cases, all animal life may disappear;

(c) change in the colour, transparency and turbidity of the water and temporary reduction of photosynthesis, of the phytoplankton and of primary production, particularly in the case of surface dumping. The seabed becomes covered with iron oxides and other metals where the dumping is carried out in estuaries and in shallow water;

(d) on the other hand, there is no evidence of any toxic effects on man from the consumption of species of fish caught in the discharge areas.

The report contains an inventory of the waste from the production of TiO_2; these wastes were classified into four major categories corresponding to the waste products discharged by factories at different stages of production. These categories were:

● insoluble matter after filtration
● 'copperas' (ferrous sulphate)
● strong acids
● weak acids or weak liquors.

The report concluded that it would seem fair to suggest that industries in this sector should, within reasonable time limits; (a) store on land the insoluble matter after filtration; (b) make certain reductions in the total pollution.

(2) Directive on waste from the industry

In the light of these findings the Commission submitted to the Council a proposal for a Council directive on waste from the titanium dioxide industry. The aim of this directive was gradually to reduce and then to eliminate pollution of the sea by waste from the titanium dioxide industry.

There were three aspects to the proposed directive, namely:

● prior authorization
● ecological control of the environment
● measures which were to be taken to reduce and eliminate pollution and nuisances.

The proposed directive provided that the competent authority of the State in whose territory the industrial establishment is located should only grant an authorization for dumping at sea or discharge into estuaries if: there was no adverse effect on boating, fishing, leisure activities, or extraction, desalination, fish and shell fish breeding, on regions of special scientific value and on other legitimate uses of the sea; and no other means of destruction or disposal existed.

The proposed directive provided that irrespective of the method and extent of treatment of effluent which was discharged, any discharge into a marine area or into an estuary should be accompanied by a systematic follow up on the general ecology of the environment. The follow up should include in particular; an ecological inventory of the current state of the area affected by the release and sampling species of molluscs, crustaceans, fish and plankton organisms.

Discharge operations should be suspended if:

(a) an examination of the general ecology of the area revealed a marked deterioration of that ecology;
(b) tests for toxicity induced by the accumulation of metals in food chains indicated hazard to human health; and
(c) the results for the tests for acute toxicity were at variance with certain values set out in the directive.

The directive envisaged a staged transition period to allow industry to adapt from the present situation to one where there was almost total elimination of dumping at sea. In the Commission's proposal:

● as from 1 January 1978, new industrial establishments would have to reduce the pollution they caused to 30% of the total untreated pollution, and as from 1 January 1985 to 5% thereof;
● existing industrial establishments would have to effect a reduction to 70% as from 1 January 1978, 30% as from 1 January 1981 and 5% as from 1 January 1985.

In the Commission's view, these percentage reductions were based on perfectly feasible techniques.

The Council discussed but failed to agree on the draft directive at its meeting on 14 June 1977. The directive was finally agreed at the Council meeting of 12 December 1977. In adopting the directive, the Council retained the two aspects of prior authorization and ecological control of the environment. The Council was, however, unable to agree to the Commission's proposal regarding the step-by-step elimination of the wastes resulting from the titanium dioxide industry. Instead, the Council agreed that Member States should draw up programmes for the progressive reduction and eventual elimination of pollution caused by waste from existing industrial establishments. These programmes should set general targets for the reduction of pollution from liquid, solid and gaseous waste to be achieved by 1 July 1987. The programmes should also contain intermediate objectives. They should, moreover, contain information on the state of the environment concerned, on measures for reducing pollution and on methods for treating waste that is directly caused by the manufacturing processes.

The programmes should be sent to the Commission by 1 July 1980 so that it might, within a period of six months after receipt of all the national programmes, submit suitable proposals to the Council for the harmonization of these programmes in regard to the reduction and eventual elimination of pollution and the improvement of the conditions of competition in the titanium dioxide industry. The Council should act on these proposals within six months of the publication of the Opinion of the European Parliament, and that of the Economic and Social Committee. Member States should introduce a programme by 1 January 1982.

The Council agreed that the programmes referred to must cover all existing industrial establishments and must set out the measures to be taken in respect of each of them. Where, in particular circumstances, a Member State considered that, in the case of an individual establishment, no additional measures were necessary to fulfil the requirements of the Directive, it should, within six months of notification of the Directive, provide the Commission with the evidence which had led it to that conclusion.

After conducting any independent verification of the evidence that may be necessary, the Commission might agree with the Member State that it was not necessary to take additional measures in respect of the individual establishment concerned. The Commission must give its agreement, with reasons, within six months.

If the Commission did not agree with the Member States, additional measures in respect of the establishment should be included in the programme of the Member State concerned.

Two member states asked the Commission for permission not to draw up pollution reduction programmes. The Commission agreed in two cases but said in the other cases programmes should be drawn up. The Commission's opinion was contested by two companies which brought an action in the Court of Justice. The case has not been settled.

(3) Surveillance of waste discharged by the titanium dioxide industry
On 30 December 1980 the Commission submitted to the Council a proposal for a directive on methods for the surveillance and monitoring of the environments affected by wastes from the titanium industry in accordance with the requirements of Article 7 of the directive agreed in December 1977. The aim of the proposal was to provide the means of identifying pollution levels in those parts of the environment affected by discharges of titanium waste (including air, water, and land), of understanding pollution trends and of assessing the progressive reduction of pollution caused by these wastes. Where waste is discharged into fresh water or dumped at sea, monitoring of certain parameters in the water column, living organisms and the sediment would be compulsory: monitoring of other parameters would be optional. Where waste is stored, tipped or injected, monitoring would include tests to ensure that surface and ground waters have not been contaminated. General rules for sampling, frequency of sampling, analysis of parameters and methods of measurement would also have to be observed. The proposal was agreed by the Council on 24 June 1982.

(c) Future work

The Commission stated in 1980 that in view of differences in the Council they saw no point in continuing activities in this sector and presenting specific proposals for other branches of industry.

(d) Thermal discharges

The Commission in a preliminary report of 3 April 1974 on the problems of pollution and nuisances originating from energy production placed a special emphasis on the problem of thermal discharges. The report noted that concern over the environmental effects of the discharge of heat into natural bodies of water was steadily mounting. An increase in the natural temperature of lakes and streams was reported to affect the growth rate and, in some cases, the species of aquatic flora and fauna. Evidence indicated that at some locations certain forms of aquatic life had benefited by increased water temperatures and that this water, used in irrigation, could help promote plant growth. In general, however, the ecological effects of increased temperature were considered detrimental, and opposition to increasing water temperatures was growing.

The report noted that the solubility of oxygen in water decreased continuously as temperature went up. On the other hand, the oxygen demand for the biological degradation processes of organic pollutants carried within the water increased with temperature, because these processes were generally accelerated by higher temperatures. Thus, discharge of heat into rivers, raising the water temperature well above its natural level, was detrimental and might endanger this natural resource and aquatic life, especially if the body was already heavily polluted by organic and inorganic matter.

The report noted that power plants constituted the most important source of thermal discharge. In the light of present technology this situation was not likely to change much within the next decade, whereas electricity production would certainly continue to grow at a similar rate as in the past decade, i.e. a doubling of the production in about 10 years. The growth rates of later decades to the end of this century were open to speculation but they were unlikely to fall below 5%.

As a consequence the total quantity of heat discharged directly into the environment by power plants might increase by at least a factor of 8 in the year 2000 as compared to the situation in 1970, even if—for the individual plant—it might be lowered by improved technology and other uses of waste heat.

Already within this decade in most Member States the cooling capacity of inland surface waters would be exhausted if the open cycle cooling technique mainly used hitherto was not substituted by the introduction, on a large scale, of wet cooling towers.

However, with increasing unit size of power plants and with a trend to installing multiple units at a given site, limits to the generalized use of wet cooling towers were imposed by the large quantities of water evaporated in the towers (effect on the microclimate) and by the water withdrawn from the rivers. If, for instance, the total evaporation from wet towers was limited to 2% of the average annual flow of European rivers, and to 10% of the flow in periods of low water coinciding with high electricity production, it would not be possible to evacuate more than 5,000 TWh* of waste heat per year by these wet towers. Under this hypothesis some of the EEC Member States would be confronted from the 1990s on with the need to look for other methods of evacuation of waste heat from electric power production.

The report suggested that this situation could be tackled in different ways, each of them presenting problems with regard to environment protection, to the rational use of natural resources, to the economy of energy production, and to the development of regional activities:

- by the installation of power plants equipped with dry cooling towers where the waste heat was discharged directly to the atmosphere;
- by the location of new multiple unit power plants at the sea shores where eventually a cheap open cooling circuit could be used;
- by the decentralization of power production in smaller plants which were equipped with dry cooling towers and the devices required for compliance with emission limits set for air pollutants, which were located near to large urban areas, and which might use the waste heat for district heating and other purposes.

The report concluded that at the present stage of knowledge a thorough

* TWh = 10^6 megawatt hours = 86,000,000,000,000 kcal

investigation of the problem related to thermal discharges must be under-taken. The studies must be directed towards:

- the establishment of plants with regard to the thermal discharges which could be directed to rivers, and especially to coastal waters, without damaging the ecosystem;
- an increased development of dry cooling towers through appropriate measures such as R & D contracts, exchange of information and experience.

In parallel to these studies, guidelines should be elaborated facilitating the siting of new power plants at carefully chosen places in consideration of the requirements of environmental protection.

These guidelines should take into account not only environmental and economic considerations but also give thought to the most efficient use of natural resources in terms of energy consumption, land use and water use.

In its resolution of 3 March 1975 on energy and the environment the Council took note of the Commission's preliminary report on the problems of pollution and nuisances relating to energy production and, where thermal discharges are concerned, called on the Commission to submit proposals on policies to be followed by the Communities and Member States on the following matters:

1. Collation of existing data on the effects of thermal discharges on the environment and further study in this field;
2. exchange of information at Community level on planning the siting of new power plants, taking into account pollution and nuisance hazards;
3. the need, wherever environmental protection so requires, to equip new power stations with cooling towers and to improve as rapidly as possible the design and technology of dry cooling towers, so as to diminish the disadvantages which the latter still present with regard to certain aspects of the environment;
4. utilization of waste heat.

5. MONITORING

Under the heading 'Exchange of information between the surveillance and monitoring networks' the Action Programme on the Environment specifies that the aims and content of Community action are:

- to organize and develop technical exchanges between the regional and national pollution surveillance and monitoring networks and to adopt all appropriate measures to improve the efficiency, accuracy and com-parative value of the devices already set up;

- to investigate, when appropriate, the desirability of setting up a system of exchanges of information on the data collected by the networks and in such cases to entrust to the Commission the analysis, for the purpose of interpretation on a Community basis, of the data collected by the national networks;
- to facilitate the inclusion of the existing networks in the Community into the framework of the global monitoring system contemplated by the United Nations.

On 1 April 1976 the Commission submitted to the Council a proposal for a Council Decision establishing a uniform procedure for the exchange of information on the quality of surface fresh water in the Community. The Commission was assisted in the preparation of the technical aspects of the decision by a group of national experts. The group, which met five times, provided information on the existing sampling stations, the parameters measured and the rivers monitored in the Member States. It advised the Commission on the list of parameters to be taken into consideration at Community level and on the criteria to be adopted for the selection of sampling or measuring stations along the principal national and international rivers.

The sampling or measuring stations, which it was proposed should form part of a Community network, were chosen on the basis of certain criteria, the most important of which were that the stations were:

- in existence and already providing information periodically;
- at points which fairly represented the conditions in the vicinity and were not subject to the direct and immediate effect of a pollution source;
- capable of assessing all the parameters considered;
- in general, not more than 100 km apart on the principal rivers, excluding tributaries;
- upstream of confluences and not below the tidal limit.

The Commission's proposal recognized that in a second stage more rivers could be added to the list in the light of experience and depending on whether new sampling or measuring stations had been set up.

The proposal also set out a list of parameters on which information was to be exchanged (see Table 2.16). The parameters selected covered the physical, chemical and microbiological properties of the water. Parameters for radioactivity were excluded from the proposal, since they are to be measured under the provisions in force in the Member States, in compliance with the basic standards set out in the Euratom Treaty.

If a meaningful comparison was to be made between the data at Community level, the information transmitted must include not only the numerical data on the parameters but also a description of the measuring methods used, of the sampling procedures, e.g. depth at which samples are

Table 2.16: Council Decision of 12 December 1977
Exchange of information in quality of surface water

Parameters in respect of which information is to be exchanged
(Modes of expression and significant figures for the parametric data)

	Parameter	Mode of expression	Significant figures	
			Before the decimal point	After the decimal point
physical	Rate of flow (^{1}) (at the time of sampling)	m³/s	× × × ×	× ×
	Temperature	°C	× ×	×
	pH	pH	× ×	×
	Conductivity at 20°C	μS cm⁻¹ at 20°C	(<100) × × (≥100) × × ×	
chemical	Chloride	Cl mg/l	(<100) × × (≥100) × × ×	
	Nitrate	NO₃ mg/l	× × ×	× ×
	Ammonia	NH₄ mg/l	× × ×	× ×
	Dissolved oxygen	O₂ mg/l	× ×	×
	BOD₅	O₂ mg/l	× × ×	×
	COD	O₂ mg/l	× × ×	×
	Total phosphorous	P mg/l	× ×	× ×
	Surfactants reacting to methylene blue	Sodium lauryl sulphate eq. mg/l	× ×	× ×
	Total cadmium	Cd mg/l	×	× × × ×
	Mercury	Hg mg/l	×	× × × ×
microbiological	Faecal coliforms	/100 ml	× × × × × ×	
	Total coliforms (^{3})	/100 ml	× × × × × ×	
	Faecal streptococci (^{2})	/100 ml	× × × × × ×	
	Salmonella (^{2})	/1 l	×	

(^{1}) The date of sampling must be given.
(^{2}) The data relating to this parameter shall be exchanged when it is measured.

taken, distance from the bank etc., and of sample preservation methods. The draft decision provided that each Member State was to designate a central agency responsible on its national territory for collecting and transmitting data to the Commission, and for receiving via the Commission the data from the other Member States. The list of central agencies is set out in an annex to the decision. The Commission will draw up and publish an annual consolidated report based on the information sent by each central agency.

The Council approved the draft decision at its meeting of 12 December 1977. A report by the Commission covering 1976 has been prepared but has not yet been published.

6. PRODUCTS WHICH POLLUTE WATER

The Action Programme on the Environment required the Commission to make investigations into the problems raised by the presence of particularly active pollutants in, amongst other things, products for the treatment of plants and animals and cleaning and conditioning agents. The programme stated that investigations should be concerned with the harmfulness, design and composition of these products, the technical possibilities of modifying their composition or of finding substitutes for them, the precautions to be taken in using them, etc. and the economic implications of the various measures under consideration.

(a) Fertilizers

The Commission's preliminary studies had shown that problems of pollutants brought into surface and ground water through agricultural practices had begun to rank alongside the better known problems of pollution due to industrial and domestic wastes.

The application of commercial fertilizers had increased dramatically in recent years, and there was much public speculation that fertilizers were contaminating drinking water and causing nutrient build up in water bodies with the consequences of undesirable growth of algae and other aquatic plants.

The question of whether increasing usage of fertilizers was contributing significantly to eutrophication and to potential health hazards was the basis of considerable controversy, and there was a general lack of sound scientific data on this subject. Systematic observation and research were needed to determine the extent to which nutrient losses might be serious. It was possible, however, that the use of fertilizers in certain localized areas and situations that were particularly vulnerable to leaching and runoff and erosion might result in nitrate and phosphate entering natural waters in undesirable quantities.

Until the role of fertilizers is well identified, as a result of scientific research, it might be advisable to impose regulations and corrective measures on the rate, time and method of fertilizer application, especially in the problem areas. There should be also a re-ordering of priorities in the fertilizer industry to produce materials which were less harmful to the environment, with particular reference to slow-released nitrogen fertilizers and nitrification inhibitors.

The Council, in the framework of the Programme for Industrial Policy Elimination of Technical Barriers to Trade, adopted on 18 December 1975 a Directive dealing with the composition, packaging and labelling of the inorganic solid fertilizers.

On 15 December 1975 the Commission presented a further proposal for a directive on the placing on the market of ammonium nitrate fertilizer, which was adopted on 15 July 1980.

(b) Detergents

On 22 November 1973 the Council adopted a directive on the approximation of the laws of the Member States relating to detergents. In proposing the directive the Commission had noted that the increasing use of detergents was one of the causes of pollution of the natural environment in general and the pollution of waters in particular. More specifically it had noted that one of the pollutant effects of detergents on waters, namely the formation of foam in large quantities, restricted contact between water and air and rendered oxygenation difficult. This caused inconvenience to navigation; it also impaired the photosynthesis necessary to the life of aquatic flora, exercised an unfavourable influence on the various stages and processes for the purification of waste waters, caused damage to waste water purification plants and constituted an indirect micro-biological risk due to the possible transference of bacteria and viruses.

The Commission further noted that the laws in force in the Member States for ensuring the biodegradability of surfactants differed from one Member State to another and this resulted in a hindrance to trade.

Under the Council directive of 22 November 1973, Member States must prohibit the placing on the market and the use of detergents where the average level of biodegradability of the surfactants contained therein is less than 90% for the anionic, cationic, non-ionic and ampholytic categories.

A further Council directive of the same date also established test procedures by which Member States should ascertain whether or not a detergent complied with the requirements laid down. If a Member State decided to prohibit a detergent, it had immediately to inform the Member State from which the product came and the Commission, stating the reasons for its decision and details of the tests undertaken. The directive provided for a mechanism of arbitration in the event of a dispute.

A further directive, relating to methods of testing the biodegradability of non-ionic surfactants, was submitted by the Commission to the Council on 8 February 1980 and was adopted on 31 March 1982 together with another amending the directive of 22 November 1973 concerning anionic detergents.

7. SEA POLLUTION

In adopting the Action Programme on the Environment the Council noted that of all the different forms of pollution, marine pollution constituted one of the most dangerous, because of the effects it had on the fundamental biological and ecological balances governing life on our planet. The danger was even greater on account of the level of pollution which had already been reached, the diversity of pollution sources and the difficulty of ensuring that any measures adopted are complied with.

The Council stated that the sea was an essential source of products and proteins, which were extremely valuable in a world which was becoming increasingly over-populated. In addition, the sea played a vital role in maintaining the natural ecological balance by supplying a large proportion of the oxygen upon which life depended. The sea and coastal areas were also of tremendous importance for recreation and leisure.

The Council noted that the pollution of the sea affected the whole Community, both because of the essential role played by the sea in the preservation and development of species and on account of the importance of sea transport for the harmonious economic development of the Community.

The Action Programme broke down the marine pollution into four main sources:

- discharge of effluents from land;
- deliberate dumping of waste at sea;
- exploitation of marine and sub-marine resources, especially exploitation of the seabed;
- sea transport and navigation.

In adopting the Second Action Programme on the Environment on 13 June 1977, the Council confirmed the general orientation of the Community's work as far as the prevention and reduction of pollution of fresh and sea water was concerned. It stated that 'a number of important provisions already adopted or contemplated constituted the basis for a coherent policy designed to prevent and reduce this type of pollution at Community level'. Action in this field during the coming years would therefore be directed towards the continued implementation of these provisions.

(a) Discharge of effluents from land

The measures for reducing land-based marine pollution have been discussed above in the content of the Council directive of 4 May 1976 concerning the reduction of pollution caused by certain dangerous substances discharged into the aquatic environment of the Community. Reference should also be made to Chapter 11 of this book—International Action—where the Community's role in the context of the Paris Convention for the Prevention of Marine Pollution arising from Land-Based Sources and the Barcelona Convention for the Protection of the Mediterranean Sea (with its protocol on Land-Based Sources) is described.

(b) Deliberate dumping of wastes at sea

As far as the deliberate dumping of wastes at sea was concerned, the Programme of Action pointed out that two major international agreements were designed to protect the marine environment from this form of pollution:

1. the Convention for the Prevention of Marine Pollution by Dumping from Ships and Aircraft, concerning the areas of the north-east Atlantic, the North Sea and their dependent seas, which was signed at Oslo on 15 February 1972 and entered into force on 7 April 1974 (Oslo Convention);
2. the Convention on the Dumping of Wastes at Sea, concerning all the seas in the world, which was signed at London on 13 November 1972 and entered into force on 31 August 1975 (London Convention).

The Council recognized that the application of international agreements in this field would necessitate the implementation within the Community of legislation and rules which would to have harmonized so as to avoid creating distortions in trade and the distribution of investments. The Council pointed out that it would be necessary to aim in particular at the application of a uniform system of licensing in the Community and to harmonize the legislation and rules concerning the dumping of substances not included in the international agreements and, if necessary, to put forward Community proposals amending the list of substances set out in the agreements.

Mention must also be made of the fact that, in the context of the Council decision on the reduction of pollution caused by the discharge of certain dangerous substances into Community waters, the Council made a specific declaration requesting the Commission to submit to it as soon as possible, and in the light of existing international conventions in this field, proposals for rules on pollution, *inter alia* in territorial waters, by operational discharges from ships and dumping from ships.

The Commission put forward a proposal for a directive on dumping wastes at sea on 12 January 1976. It was intended to define a common field of action within which rules on dumping were to be applied. The notion of 'dumping' was defined and obligations would be imposed on Member States to take appropriate measures to prevent and abate pollution of the sea thus caused.

The directive drew a distinction between three categories of waste; particularly harmful substances, the dumping of which is always prohibited (Annex I) and other harmful waste and matters, whose discharge into the sea would require the granting of a special permit by the competent national authorities (Annex II). It also laid down that dumping of all other wastes will require a prior general permit from the competent authorities.

Mention must be made of the fact that the lists of harmful substances as set out in the proposal, are also to be found in the annexes to the Barcelona Protocol which, being the latest of the above mentioned international agreements, takes into account the results of the most advanced research in this field. The lists therefore are somewhat different from those of the Conventions of Oslo and London which in addition are different from each other.

At its meeting on 19 December 1978, the Council authorized the Commission to enter into negotiations leading to the eventual accession by the Community to the Oslo Convention. The representatives of the Commission indicated that the Commission would be prepared to withdraw its proposed directive once the Community's participation in the Oslo Convention had been achieved. Chapter 11, International Action, gives further details about the Commission's proposal concerning accession to the Oslo Convention.

(c) Exploitation of marine and submarine resources, especially exploitation of the seabed

On 9 June 1977, the Commission sent a Communication to the Council on measures for the prevention, control and reduction of pollution caused by accidental discharges of hydrocarbons into the sea, accompanied by a draft Council Resolution on the same subject. The Commission Communication referred to the uncontrolled blow-out of gas and oil at 2330 hours on 22 April 1977 following a technical accident on the Bravo drilling rig. This rig lay in the Ekofisk field in that area of the North Sea where the Norwegian authorities are responsible for working.

The incident occurred when a safety valve was being attached to the top of a drilling pipe. The blow-out was halted at 1200 hours on 30 April. According to the estimates available, 20,000 tonnes of oil escaped.

In the Commission's view, the Ekofisk accident underlined the need for a more effective policy. The Commission was of the opinion that the Community should have powers and means allowing it to take effective action in such situations, so expressing the solidarity of the Member States among themselves and towards non-member countries hit by a disaster of this type, since the protection of the seas was a duty for all the countries of the international community in the interest of future generations.

The following measures should be implemented as soon as possible:

- The establishment of a 'data bank' at Community level recording the means available for taking action in the event of accidental discharges of hydrocarbons.
- The development of a research programme into the technologies for collecting and dispersing hydrocarbons, into what happens to hydrocarbons in the sea and into their effects on marine fauna and flora.

The Commission also intended to appoint a group of high-level experts to examine the causes, circumstances and effects of recent accidents involv-

ing considerable spills of hydrocarbons in the sea, the remedies and preventive measures. The group would also be responsible for studying
preparations for an environmental impact report on installations for prospecting and drilling for hydrocarbons at sea. It could also prepare the
formulation of Community positions aiming at the efficient implementation
of international agreements on:

1. the prevention of marine pollution in the prospecting and drilling for
   hydrocarbons;
2. the civil liability for damage caused by pollution with hydrocarbons
   and the establishment of a compensation fund.

The Council of 14 June 1977 was not in a position to discuss the Commission's Communication or draft Resolution. However, many of the ideas
contained in the Communication and Resolution were taken up again in
the programme of action presented to the Council by the Commission
following the Amoco Cadiz disaster. (See below.)

(d) Sea transport and navigation

(1) The Amoco Cadiz disaster
The ecological, social and economic consequences of the *Amoco Cadiz*
accident had made the general public sharply aware of the absence of
effective measures against marine pollution caused by oil tankers. At the
meeting of the External Affairs Council on 4 April 1978, Mr. Guiringaud,
the French Minister, had appealed to the Community to implement a
number of practical measures.
    At its meeting in Copenhagen on 7 and 8 April the European Council
decided that the Community should make the prevention and combating
of marine pollution, particularly by hydrocarbons, a major objective. In
the words of its final communiqué the European Council 'invited the
Council, acting on proposals from the Commission, and the Member States
forthwith to take appropriate measures within the Community and to adopt
common attitudes in the competent international bodies concerning in
particular:

(a) the swift implementation of existing international rules, in particular
    those regarding minimum standards for the operation of ships;
(b) the prevention of accidents through coordinated action by the Mem-
    ber States,
    ● with regard to a satisfactory functioning of the system of com-
      pulsory shipping lanes,
    ● and with regard to more effective control over vessels which do
      not meet the standards;
(c) the search for and implementation of effective measures to combat
    pollution.'

The Commission in response sent a Communication to the Council on 27 April 1978 putting forward a number of proposals for action by the Community.

(2) Adoption of Action Programme

At its meeting of 28 June 1978, the Foreign Ministers' Council adopted an action programme for the European Communities on the control and reduction of pollution caused by oil spills at sea. The Commission would undertake the preliminary studies necessary for the placing of appropriate proposals before the Council as soon as possible with a view to the following measures:

(a) Computer processing of the existing data, or data still to be collected, on ways of dealing with marine pollution by oil with a view to the immediate use of these data in the event of accidental pollution.

The study undertaken by the Commission would relate to any further steps to be taken so as to ensure that Member States, and if necessary interested third countries, have complete, rapid and reliable information on ways of dealing with marine pollution by oil. The information studied would include in particular:

● the teams of qualified personnel to combat marine pollution by oil, the specific qualifications of such teams and practical details for calling on their services;

● the availability of products and equipment for the chemical treatment of oil spills (dispersants, precipitating agents, aerator-agitators and dispersant spreading equipment);

● availability of equipment for the mechanical treatment of spills and the specifications of the ships and equipment from which these operations can be carried out (deployment, speed and stability of vessels, equipment fitted) and their location;

● availability of the requisite resources to protect coastline (oil slick containment equipment, oil absorbents or precipitating agents, gels enabling dispersants to be used on rocks or structures, aerator-agitator and recovery equipment for use in shallow waters and beach-cleaning equipment).

The study would consider in particular the conditions in which the collection and computer processing of this information, which should be constantly updated, could be achieved at least cost while avoiding all duplication. Possibilities which already existed both at national level and within the framework of existing conventions should be fully explored.

(b) Study of the availability for the Member States of relevant data on tankers liable to pollute the Community's waters or coast and on man-made structures under the jurisdiction of Member States.

The study undertaken by the Commission would be aimed at:

(1) investigating to what extent Member States may already have

rapidly available complete and reliable information relating to:
- the structural features of the tankers and artificial structures in question and emergency plans for prompt intervention;
- any infringements committed in the territorial waters.

(2) investigating, if necessary, the means of improving the availability of this information.

(c) Study of the need for measures to enhance the cooperation and effectiveness of the emergency teams which have been or are to be set up in the Member States.

Effective pollution control implied the existence of fully and suitably trained specialist groups supplied with equipment which was easy to use and immediately available if required.

The Commission would study the need to improve cooperation between States on a plurinational and, if necessary, Community basis. The study would deal in particular with training methods, the compatibility of the various material and equipment used and training programmes. The Commission would put to the Council proposals to harmonize the technical characteristics of the various equipment used, if appropriate.

(d) Study of a possible Community contribution to the design and development of clean-up vessels to which may be fitted the equipment needed for the effective treatment of oil spills.

The treatment of oil spills required the deployment of a range of equipment to be used as the circumstances dictate. It had been suggested that all this equipment be assembled on clean-up vessels fast enough to carry it promptly to the site of an emergency. Design and development costs of such vessels seemed to be substantial. The Commission would study the merits of such an operation, and the financial problems posed, as well as the possibilities open to the Community for contributing to a solution of such problems, for example through investment premiums or interest-relief grants.

(e) Study of the amendments and improvements which may have to be made to the rules of law regarding insurance against the risks of accidental pollution from oil spills.

In the case of accidents giving rise to serious pollution, the Commission would study the steps necessary to ensure a more efficient application of the 'polluter pays' principle whereby any natural or legal person governed by public or private law who is responsible for an act of pollution must pay the cost of the measures needed to prevent or control such pollution, in accordance with the Commission's communication to the Council annexed to the Council Recommendation of 3 March 1975 regarding cost allocation and action by public authorities on environmental matters.

It would in particular be necessary to study the steps taken so that compensation covers the totality of the damage suffered as a result of accidental pollution.

(f) Establishment of a proposal for a research programme on chemical
     and mechanical means of combating pollution due to oil discharged
     at sea, what becomes of it and its effects on marine flora and fauna.
         The Commission would undertake a study of the research pro-
     grammes under way at national and international level on the themes
     noted above. It would submit to the Council appropriate proposals
     with a view to completing, on this point, the Community pro-
     gramme of environmental research.

(3) Community information system

As a result of the studies carried out under the Action Programme, the
Commission submitted to the Council a proposal for a draft decision for
setting up a Community information system for the control and reduction
of pollution caused by hydrocarbons discharged at sea. The decision was
adopted on 3 December 1981 and provides for:

(a) an inventory of the means for combatting such pollution;
(b) a list of national and joint contingency plans;
(c) a compendium of hydrocarbon properties and their behaviour and
     of methods of treatment of mixtures of water-hydrocarbon solid
     matter recovered from the seas or along the coast.

The Commission will be responsible for implementing the system on the
basis of information from member states. A report will be drawn up by the
Commission every two years.

(4) Actions on international conventions

Meeting in Bonn on 6 and 7 July 1978, the European Council, (composed
of Heads of State or of Government) reaffirmed the necessity to intensify
efforts to prevent and control pollution of the sea, especially by hydrocarbons. In its Communication of 27 April 1978, the Commission also asked
the Council to agree to an accelerated ratification by Member States of the
SOLAS, MARPOL and ILO Maritime Conventions. This proposal was
agreed by the Council on 12 June 1978.
    The Commission also proposed that the Council:

(a) should approve the accession of the European Economic Community
     to the Protocol to the Barcelona Convention for the protection of
     the Mediterranean Sea against pollution, of February 1976, concern-
     ing cooperation in combating pollution by oil and other harmful
     substances in cases of emergency, and
(b) should authorize the Commission to open negotiations with the
     States party to the Bonn Agreement of 9 June 1969 on cooperation
     in dealing with pollution of the North Sea by oil, with a view to the
     Community's accession to that agreement.

In the Commission's view, the Community could play a vital role under
these agreements. It put forward proposals for a draft directive on port

state enforcement. Further action on this and the proposal on an oil tanker file put forward as part of the Community information system proposal (see (3) above) was pursued in the context of an initiative for a wider international agreement. For further details see Chapter 11, International Action, Sections 3 and 8.

8. FUTURE WORK

In its proposals for the Third Action Programme the Commission proposes to continue with the measures begun under the first two programmes and will implement the directives and decisions adopted by the Council with a view to preventing and reducing water pollution. The following are the main areas concerned:

(a) The control of pollution by dangerous substances. Future action will involve the careful selection of priorities and the use of simplified procedures, especially with regard to the substances in List I of the discharge of dangerous subtances directive. Particular attention will be paid to harmonizing programmes for reducing pollution by the substances recorded in List II of the same Directive. To make this action more effective, the Commission intends to review measures for reducing indirect or dispersed discharges of certain substances and, if necessary, will send the appropriate proposals to the Council.

(b) The control of pollution from oil spills. To combat hydrocarbon pollution effectively, it is first of all essential to accelerate the implementation of a preventive policy. The Commission proposes therefore to take steps to ensure that the international conventions concluded under the auspices of IMCO and the ILO are quickly applied, and that the opportunities afforded by these conventions are properly used. It will also act to bring into service the information system adopted by the Council while completing the study of the problems mentioned in its Communication to the Council of 26 June 1980 and will make the relevant proposals. The Commission will coordinate its work with that carried out under the Barcelona Convention on the Protection of the Mediterranean Sea against Pollution and the Bonn Agreement on Pollution of the North Sea. During this preparatory work, the Commission will regularly consult the Advisory Committee on the control and reduction of pollution caused by hydrocarbons discharged at sea which it set up on 25 June 1980.

(c) Monitoring and control with a view to improving water quality and reducing pollution.

(d) Participation in international conventions. These include the International Commission for the Protection of the Rhine against pollution, and measures aimed at protection of the North Sea and of the Mediterranean.

3

Air

1. POLLUTION BY SULPHUR DIOXIDE

(a) Criteria and air quality standards

Part II, Title I, Chapter 1 of the Communities' Environment Programme dealt with the objective evaluation of risks to human health and the environment from pollution. Priority was given to so-called 'first category' pollutants, which were chosen on the grounds both of their toxicity and of the current state of knowledge of their significance in the health and ecological fields. The Programme states that, 'in the light of this objective evaluation, it will be possible, without having resort to arbitrary values, to set limits to the presence of these pollutants in the environment and determine quality values for products in terms of standards designed to protect human health and the environment'. Sulphur dioxide and suspended particulate matter in the atmosphere were considered as first category pollutants for which action was required because of their toxicity, synergistic effects and the current state of knowledge regarding their significance for public health.

On 19 February 1976 the Commission submitted to the Council a proposal for a Council resolution concerning the determination of criteria for sulphur dioxide and suspended particulate matter in urban atmospheres. In the Commission's view the two pollutants were among those substances for which an objective evaluation of the scientific data available could be carried out with the certainty necessary for the development of criteria for the public health point of view. (In the terminology of the Communities' Environment Protection Programme 'criterion' signifies the 'relationship between the exposure of the target to pollution or nuisance, and the risk and/or the magnitude of the adverse or undesirable effect resulting from the exposure in given circumstances'.)

On the basis of these criteria, the Commission also proposed a Council directive concerning health protection standards for sulphur dioxide and suspended particulate matter in urban atmospheres.

The directive defined the meaning of air quality standards for sulphur dioxide and suspended particulates and prescribed the levels which must not be exceeded.

It required the Member States to take the necessary measures to ensure compliance with the air quality standards. It provided for exceptions, which were clearly defined, to the implementation of this directive during a transition period ending in 1987 in order to allow the Member States time to take all the measures required.

The Member States would always be able to impose more severe air quality standards and anticipate the deadlines laid down provided that these standards were not an obstacle to the proper functioning of the Common Market. The directive further laid down the procedure by which the Member States were to inform the Commission of existing pollution levels and also of proposed measures to reduce these levels.

It was emphasized that the implementation of this directive must not lead to a deterioration of air quality in the 'clean' regions; as far as possible, compliance with the standards must be achieved by reducing emissions and not by wider dispersal of pollutants in the environment.

Finally the directive laid down reference methods for analysis of the pollutants, but at the same time allowed the Member States the option of using equivalent methods and required the Commission to help Member States to demonstrate that such methods are equivalent. The directive was adopted on 15 July 1980. It fixed limit values and guide values for sulphur dioxide and suspended particulates in the atmosphere and the conditions for their application in order to improve the protection of human health and the protection of the environment. Details of the limit and guide values are given in Tables 3.1 and 3.2. The limit values which are not to be exceeded throughout the territory of member states are intended to protect human health in particular. The guide values are intended to serve as long-term precautions for health and the environment and reference points for the establishment of specific schemes within zones determined by member states. Member states are to ensure that the limit values are not exceeded from 1 April 1983 but provision is made for exceptions to be notified to the Commission together with plans for the progressive improvement to meet the limit values as soon as possible. The latest date is set at 1 April 1993. The directive provides rules for sampling and analysis, but the Commission is required to report within 8 years of adoption on the results of measurements which do not follow the prescribed method in the directive. There is also provision for consultation in border areas where specific zones are determined with lower values than the prescribed limit values.

The Council also adopted at the same time a resolution on transboundary air pollution, in which member states resolved to endeavour to limit and as far as possible gradually to reduce and prevent transboundary air pollution by sulphur dioxide and suspended particulates.

(b) Monitoring

As far as air surveillance and monitoring networks are concerned, a first step was taken when on 24 June 1975 the Council adopted a decision

Table 3.1: Council Directive of 15 July 1980
Air quality for sulphur dioxide suspended particulates

Limit values for sulphur dioxide expressed in µg/m³ with the associated values for suspended particulates (as measured by the black-smoke method (^{1})) expressed in µg/m³

Reference period	Limit value for sulphur dioxide	Associated value for suspended particulates
Year	80 (median of daily mean values taken throughout the year)	>40 (median of daily mean values taken throughout the year)
	120 (median of daily mean values taken throughout the year)	≤40 (median of daily mean values taken throughout the year)
Winter (1 October to 31 March)	130 (median of daily mean values taken throughout the winter)	>60 (median of daily mean values taken throughout the winter)
	180 (median of daily mean values taken throughout the winter)	≤ (median of daily mean values taken throughout the winter)
Year (made up of units of measuring periods of 24 hours)	250 (^{2}) (98 percentile of all daily mean values taken throughout the year)	>150 (98 percentile of all daily mean values taken throughout the year)
	350 (^{2}) (98 percentile of all daily mean values taken throughout the year)	≤150 (98 percentile of all daily mean values taken throughout the year)

Limit values for suspended particulates (as measured by the black-smoke method (^{1})) expressed in µg/m³

Reference period	Limit value for suspended particulates
Year	80 (median of daily mean values taken throughout the year)
Winter (1 October to 31 March)	130 (median of daily mean values taken throughout the winter)
Year (made up of units of measuring periods of 24 hours)	250 (^{2}) (98 percentile of all daily mean values taken throughout the year)

(^{1}) The results of the measurements of black smoke taken by the OECD method have been converted into gravimetric units as described by the OECD.

(^{2}) Member States must take all appropriate steps to ensure that this value is not exceeded for more than three consecutive days. Moreover, Member States must endeavour to prevent and to reduce any such instances in which this value has been exceeded.

Table 3.2: Guide values for sulphur dioxide and suspended particulates

Guide values for sulphur dioxide expressed in μg/m³

Reference period	Guide value for sulphur dioxide
Year	40 to 60 (arithmetic mean of daily mean values taken throughout the year)
24 hours	100 to 150 (daily mean value)

Guide values for suspended particulates (as measured by the black-smoke method (¹)) expressed in μg/m³

Reference period	Guide value for suspended particulates
Year	40 to 60 (arithmetic mean of daily mean values taken throughout the year)
24 hours	100 to 150 (daily mean value)

(¹) The results of the measurements of black smoke taken by the OECD method have been converted into gravimetric units as described by the OECD.

establishing a common procedure for the exchange of information between the surveillance and monitoring networks based on data relating to atmospheric pollution caused by sulphur compounds and suspended particulates (SPM).

Member States had six months after the adoption of the decisions to select from existing or planned sampling or monitoring stations those which are to supply the data for the exchange of information. The selection of sampling or monitoring stations was to be based mainly on geographic and demographic parameters (urban and rural areas, size of cities, residential or predominantly industrial zones) and on pollution levels (maximum, average and minimum).

The decision provided that the daily average concentrations of certain sulphur compounds and suspended particulates should be transmitted monthly to the Commission and that an annual report, to include different types of data evaluation, should be prepared by the Commission, in consultation with national experts, on the basis of this data and of further information deemed appropriate by Member States and made available to the Commission.

The Commission undertook, on the basis of current studies on the comparability of the measurement methods, to submit proposals at the earliest opportunity on the harmonization of these methods so that the data obtained by the various stations referred to in the decision might be directly compared.

Subsequent consultations between the Commission and Member States have permitted the:

- clarification and provision of a uniform interpretation for the parameters defined in the Decision and thus assisted the Members States to choose in a homogeneous way the participating stations;
- clarification of the structure of the network;
- discussion of the problems existing at the national level for the transmission of data;
- examination of the possible further uses which can be made of this network.

At the present stage the information provided by Member States will provide input into a network composed of four parts:

- high impact areas;
- open country situations;
- pilot cities, and
- comparison stations.

The annual reports covering the period 1976, 1977 and 1978 have been published.

A further proposal for extending the exchange of information on air pollution was put forward by the Commission on 26 July 1981. In order to maintain continuity with the present system the selected stations supplying data on sulphur dioxide and suspended particulates are to be transferred to a new reciprocal exchange. In addition the proposal provides for inclusion of information on lead and other heavy metal particulates, nitrogen oxides, carbon monoxide and ozone. Selection of station is to be by member states and should reflect the different types of urbanization, topography and climatology, as well as pollution levels, prevailing in the territory of the member state concerned. The Commission will organise intercomparison programmes with the co-operation of member states. Annual reports are to be produced by the Commission. The decision was agreed by the Council on 24 June 1982.

(c) Reducing pollution at source

In the report submitted to the Council on 3 April 1974 on the problems of pollution and nuisances originating from energy production the Commission stated that SO_2 emissions from stationary combustion sources were likely to pass from 14 million tonnes in 1968 to some 21 million tonnes in 1980 for the nine Member States of the Community. Though there were certain gaps in the information Member States were able to give on the expected repartition in 1980 between solid, liquid and gaseous fuels, existing data indicated that more than 50% of total consumption of stationary combustion sources in 1980 might be accounted for by liquid fuels and another 20% at least by gas. Taking into account the sulphur content of the three types of fuel, it followed that the combustion of oil would constitute the major source of SO_2 in 1980, as far as stationary combustion sources were concerned.

(1) Gas oil

The report noted, that within large urban areas SO_2 ground level concentrations were known to be strongly dependent upon the amount of emissions at low level (e.g. above all those originating in domestic heating and from small commercial installations fuelled mainly with gas oil). Within the past decade the consumption of gas oil for domestic heating purposes had increased very sharply, replacing coal. This trend was likely to continue in the near future although at somewhat smaller growth rates (because of the higher costs of oil products, the competition of natural gas and of heating by electricity).The Commission therefore prepared, and submitted to the Council, on 11 February 1974, a proposal for a directive on the approximation of the laws of the Member States relating to the sulphur content of certain liquid fuels.

The aim of the directive was to lay down Community requirements for the limitation of the sulphur content of gas oils which would have the effect of ensuring a marked reduction in atmospheric pollution caused by sulphur compounds resulting from gas oil combustion.

Having regard to the sulphur content of gas oils, only two grades of such oil would be permitted on the Community domestic market as from 1 October 1976.

With a view to ensuring, over the years to come and in spite of a sharp and continuing increase in consumption, the attainment of the objectives (reduction in the existing level of pollution, or in certain areas, maintenance of existing levels where these were deemed acceptable), successive reductions in sulphur content were envisaged for the two grades of gas oil.

As from 1 October 1976 the maximum basic type sulphur content would be restricted to 0.5% and would subsequently be reduced to 0.3%, commencing 1 October 1980.

As from 1 October 1976 the maximum content for sulphur of the second type (type B) would be limited to 0.8% and this figure, in turn, would be reduced to 0.5% as from 1 October 1980.

Thus, on and after this last-mentioned date, there would be two grades of gas oil on the market with a maximum sulphur content of 0.3% and 0.5% respectively.

In the Commission's view it seemed advisable in the present situation to leave to the Member States the task of defining the low atmospheric pollution regions and also any region where the contribution made by gas oils to atmospheric pollution was low (in these regions it would be possible to use type B gas oil).

This possibility offered a certain flexibility, particularly in the Member States which had industrial regions where competitive conditions were not favourable, but where the question of air pollution was not a pre-eminent consideration.

The Commission believed that the obligation of Member States to notify it of the criteria which they had established for determining these regions, constituted a serious curb to the excessive use of this type of gas oil.

Nevertheless, the Commission would re-examine the appropriate provisions of the directive in the light of progress accomplished in defining criteria for the harmful properties of pollutants and targets for air quality. It would also consider new data for the levels of sulphur dioxide and, if necessary, put forward appropriate proposals before 1 October 1980.

This meant that the Commission would be able to verify if the choice of criteria, and of their application, ensured a protection of the environment in these regions compatible with the Commission's programme on the subject.

In addition, so as not to impede the national programmes aimed at reducing atmospheric pollution caused by sulphur compounds, Member States would be able to bring forward the proposed dates for the reduction of the maximum sulphur content of gas oils and for the introduction of the directive.

In presenting this proposal, the Commission recognized that the production of gas oils conforming to the proposed restrictions, particularly to those foreseen for 1980, would henceforth necessitate a very large quantity of crude oil being treated in desulphurizing plants, which would call for supplementary investment to increase the capacity of these plants. To absorb this and the resultant higher production costs would call for an increase in the price of gas oil. It was difficult to establish, in the current state of the oil sector, precise estimates of the increase in costs, but the percentage was likely to be less than 5% if the trend towards considerably higher crude oil prices continued up to the date of application of the proposed measures.

The increase in the amount of crude oil to be treated similarly would lead to a greater consumption of energy by the desulphurizing plants, thereby increasing their need for crude oil. In this regard also, little data were available, but the estimates were indicative of a value less than 0.5%.

The Commission was, however, of the opinion that the forecast increases, both of the price of gas oil and of the use of crude oil, were largely justified by the basic aims of the proposed directive.

The Council adopted the Commission's proposal with certain modifications on 24 November 1975.

The Council agreed that should environmental requirements or the state of desulphurizing technology change appreciably or should the economic situation in the Community as regards the supply of crude oil change substantially, the Commission might on its own initiative or at the request of a Member State propose amendments to the sulphur content relating to the period beginning 1 October 1980. The Council might decide on such amendments by a qualified majority not later than 1 October 1977.

The Council also agreed that if, because of sudden change in crude oil supplies, changes should occur in the sulphur content of the oil such as to jeopardize supplies to consumers in view of the shortage of available desulphurization capacity, a Member State might allow onto its territory gas oils which did not conform to the specifications laid down. It should forthwith notify the Commission, which should, after consulting the other Member States, decide within three months on the duration and details of the derogation.

The Council recognized that the second stage of the programme for reducing the sulphur content of gas oil raised particular technical and economic problems for Ireland and therefore granted Ireland a five year exemption before it passed on to the second stage.

(2) Residual fuel oil

Stationary combustion installations, like boilers, furnaces, incinerators, torches etc., are estimated to create between 70 and 90% of total SO_2 and particulate emissions. After considering the availability of different types of crude, the Commission calculated in the above-mentioned report on the problems of pollution and nuisances originating from energy production that SO_2 emissions from the combustion of residual fuel oils were likely to increase markedly from 5.4 million tonnes (1971) to more than 17 million tonnes (1980) if no abatement measures were taken. In the light of these numbers, the Commission concluded that all means suited to abate SO_2 emissions from installations burning residual fuel oils (mainly large stationary ones) deserved special attention.

The European Parliament in its resolution on the Commission's preliminary report, approved this orientation, while considering that the desulphurization of fuels should be a first priority. The Council, in its resolution of 3 March 1975 on energy and environment, specifically invited the Commission to submit proposals for regulations on the sulphur content and use of heavy fuel oils.

On 19 December 1975, the Commission submitted to the Council a proposal for a Council directive relating to the use of fuel oils with the aim of decreasing sulphurous emissions.

In submitting this proposal, the Commission noted that some Member States had already taken or had proposed legislative provisions for the type of fuel oil to be used in certain regions or within certain installations, or for a limitation of sulphur content of such fuels, or for a limitation of emissions with a view to protecting public health against dangerous concentrations of sulphur dioxide, suspended particulate matter and other pollutants in the ambient air.

The proposal met a number of difficulties and, after a preliminary exchange of views between Environment Ministers at their meeting on 14 December 1977, no further progress was made. The Commission announced in March 1982 that it had withdrawn the proposal. The drop in consumption of heavy fuel oil owing to the rise in costs of energy had made the forecasts out of date and the proposal was less relevant to present circumstances. Efforts would continue to reduce oil consumption in the Community by means of the rational use of energy programme, nuclear power and wider use of coal.

2. POLLUTION BY LEAD

(a) Criteria, biological standards and screening of the population

Under the Action Programme of 22 November 1973, lead and its compounds were considered as pollutants for top priority consideration. On 16 April 1975 the Commission presented certain information on the toxic effects of lead on man taking account of industrial medical experience, cases of accidental and chronic lead poisoning and of extrapolations from animal experiments. Subsequently, the Commission proposed a draft Council directive on biological standards for lead and on screening of the population for lead.

In the Commission's view, it has been shown that lead may have a specific effect on the lungs if fairly large quantities of lead are inhaled. Metabolic studies and epidemiological surveys had also shown that it was possible to establish a relationship between the amount of lead inhaled and the increase in blood lead level.

Since atmospheric pollution by lead is more easily controlled than lead pollution from other sources, it was desirable to limit as much as possible the contribution of inhaled lead to the total blood lead level and to set a maximum contribution of lead in blood.

On 29 March 1977, the Council adopted the Directive on biological screening of the population for lead. Under the Directive, sampling of blood lead levels was to be carried out over a four year period by Member States on:

- groups of at least 100 persons in urban areas with more than 500,000 inhabitants;
- groups of at least 100 persons, insofar as this was feasible, chosen from among people exposed to significant sources of lead pollution;

● critical groups determined by the competent authorities in the Member States.

In each Member State and during each campaign the number of analyses to be performed should be 50 or more per million inhabitants. All blood sampling was to be carried out on volunteers.

The sampling of the groups referred to above should be carried out during at least two campaigns in each area investigated during the period of operation of the programme, separated by at least 24 months. In the second campaign, samples should not necessarily be taken from the same persons as in the first campaign.

In assessing the results of the biological screening the following blood lead levels should be taken, together, as reference levels:

● a maximum of 20 μg of Pb/100 ml of blood for 20% of the group of people examined;
● a maximum of 30 μg of Pb/100 ml of blood for 90% of the group of people examined;
● a maximum of 35 μg of Pb/100 ml of blood for 98% of the group of people examined.

Where the results of the analyses indicated that the reference levels set out in Article 6 had been exceeded in one or more cases Member States should:

● check the validity of the results;
● take action to trace the exposure sources responsible for the levels being exceeded; this should also include action on all individuals with a blood lead level over 35 μg/100 ml;
● take all appropriate measures at the discretion of their competent national authorities.

Within six months of notification of the Directive, the Member States should designate the competent national authorities which should forward to the Commission:

● the data relating to the biological screening of the population groups referred to, together with details of the methods of analysis, the population groups examined and the areas in which samples have been taken; complete anonymity should be preserved as regards the persons examined; the Commission and the Member States should agree on the procedures and the method whereby these data should be forwarded;
● information on the causes or factors presumed to have resulted in the reference levels being exceeded and measures taken by the competent national authorities.

At least twice a year the Commission should convene a meeting of representatives of the Governments of the Member States, which should, in particular:

- ensure that implementation of the biological screening is harmonized;
- see that the analyses carried out are comparable;
- examine the information and facilitate the exchange of information between the Member States on the results of the biological screening and on the measures taken.

On the basis of the information collected, the Commission should draw up in cooperation with the competent national authorities:

- a collated annual report on the implementation of the programme, which should be forwarded to the Member States, the Council and the European Parliament;
- a general report at the end of the programme which would form the basis for drawing up any further proposals taking account of progress made in scientific and technical knowledge.

The first survey took place in 1979. A report on its results showed that a total of close to 18,000 subjects, evenly distributed between men and women, were examined throughout the Community. A significant percentage of the subjects involved were children. 168 separate areas and population groups were investigated. Preliminary conclusions were that:

(a) the overall campaign had been monitored by a very comprehensive quality assurance programme lending confidence to the interpretation of results;

(b) in general blood lead levels for the population of the Community were lower than could have been anticipated from earlier fragmentary studies;

(c) the studies conducted in urban areas with no known specific sources of lead (point emission sources from industry, plumbosolvency and lead pipes) seemed to indicate that there was no special risk for the population taking into account the reference values provided for in the directive;

(d) studies conducted in areas with known specific sources and in particular for critical population groups such as children of lead workers or children living near lead works, confirmed that a health risk might exist since in some of those instances reference levels were exceeded. (This was the case for known problem areas and for newly found potential problem areas.)

Member States are taking active measures to circumscribe the problem in a better way; remedial action is being taken where necessary. In the second campaign in 1982 special emphasis is being placed on such types of areas.

(b) Air quality standards

The Commission also proposed another directive whose purpose was to establish air quality standards for lead to be respected by Member States in order to protect their populations' health from the effects of atmospheric pollution by lead outside the place of work.

For the purpose of this directive air quality standards meant the maximum atmospheric concentrations of lead at which lead has no specific effect on the lungs and which ensure that the contribution of atmospheric lead to the total body burden may be kept down to less than one quarter.

The air quality standards proposed were the following:

1. an annual mean level of not more than 2 micrograms Pb/m^3 in urban residential areas and areas exposed to sources of atmospheric lead other than motor vehicle traffic;
2. a monthly median level of not more than 8 micrograms Pb/m^3 in areas particularly exposed to motor vehicle traffic.

Member States were to take the necessary measures to ensure that these standards were respected by 1980; such measures might in no event lead to an increase in atmospheric lead in low-concentration areas or an increase in the concentration of other atmospheric pollutants.

The directive proposed certain monitoring procedures for measuring the level of atmospheric lead pollution.

The Commission's proposal was agreed by the Council on 24 June 1982.

As adopted the directive sets a single limit value of 2 micrograms of lead per cubic metre specifically to help protect human health against the effects of lead in the environment. Member States are required to respect this value within five years of notification. Sampling stations are to be set up where individuals may be exposed continually for a long period and where the limit values are likely not to be respected. Member States are required to notify the Commission four years after notification of places where the limit value is exceeded and to submit plans for progressively reducing concentration to the required limit within seven years from notification.

(c) Problem of lead in petrol

The lead emissions from motor vehicles constitute a large part of the total quantity of this element in the atmosphere, in particular in the atmosphere of our big cities. Early in the 1970s, it became apparent to the Commission that the differences between the laws, regulations or administrative measures in the EEC Member States on the limitation of the lead content of petrol for motor vehicles threatened to create obstacles to the free movement both of fuels and motor-vehicles within the Community.

These differing regulations not only caused the user trouble when travelling in the Member States but also increased his expenses, since the oil and motor industries were obliged specially to produce and export products which complied with the different regulations of individual countries. This problem was particularly acute in the motor industries, where profitability depended on large production runs.

For these reasons, the Commission on 5 December 1973 submitted to the Council a proposal for a Council directive concerning the approximation of the laws of the Member States relating to the composition of petrol.

This directive, as an initial stage, provided for a reduction of the lead content of all qualities of petrol to 0.4 g/l as from 1 January 1976. This reduction represented a considerable step forward for the Community as a whole, having regard to the laws then in force in certain Member States.

The Commission also considered it appropriate to fix, as a guide, a limiting value of 0.15 g/l as from 1 January 1978 for regular-quality petrol, mainly as a pointer for the longer term policy in the oil industry in deciding future petrol blends.

At their meeting on 30 May 1978, EEC environment ministers adopted the proposed Council Directive. In its final version, the Directive provided that from 1 January 1981, the maximum permitted lead compound content, calculated in terms of lead, of petrol placed upon the Community internal market should be 0.40 g/l. However, a Member State might require, in respect of petrol placed upon its market, that the maximum permitted lead content be less than 0.40 g/l. It should not establish limits lower than 0.15g/l.

Member States should take all appropriate steps to ensure that the reduction of the lead content does not cause a significant increase in the quantities of other pollutants or to a deterioration in the quality of petrol.

Where a Member State established, on the basis of a test performed in accordance with the procedures laid down in the Directive, that any petrol failed to comply with the requirements of Articles 2 and 3, it should take the necessary measures to ensure that these requirements are fulfilled.

Member States should, at the Commission's request, supply it with information on:

(a) the effects of the implementation of the Directive
(b) developments as regards systems to reduce the emission of lead and of any polluting substitutes in exhaust gases;
(c) development of the concentrations of lead and polluting substitutes in the urban atmosphere and their effect on public health;
(d) the effects on energy policy of the various possible ways and means of reducing pollution caused by lead emission in exhaust gases.

The Commission should report to the Council and the European Parliament on the information thus obtained and, in the light of the data compiled, should make any suitable proposals for such data to be taken into account in order to develop further Community policy on the lead content of petrol.

In a declaration made when the Council adopted the Directive, the Commission indicated that in making such further proposals, it would take into account not only the need to reduce the emission of different atmospheric pollutants, especially lead, but also of the need to reduce as far as possible the differences in the levels of lead admissible in the Member States. It was of the view that the present Directive was only a first step towards the approximation of legislation in this field.

The Directive as adopted by the Council provided that the Government of Ireland might for a period of five years commencing on 1 January 1981, permit petrol to be placed on the market in Ireland even though its lead content is greater than 0.40 g/l, without, however, exceeding the current content of 0.64 g/l.

Before the end of the five-year period the Council should, acting by a qualified majority on a proposal from the Commission, decide on the duration of a second derogation period of not more than five years.

The Directive also provided that if, as the result of a sudden change in the supply of crude oil or petroleum products, it became difficult for a Member State to apply the limit to the concentration of lead in petrol, that Member State might, after having informed the Commission, authorize a higher limit within its territory for a period of four months. The Council, acting by a qualified majority on a proposal from the Commission, might extend this period.

At the beginning of 1982, the European Parliament's Committee on the Environment, Public Health and Consumer Protection was considering a motion tabled on 24 September 1981 by Mr Collins and Mr Seefeld on behalf of the Socialist Group which asked the Commission to take action to achieve lead-free petrol by the progressive reduction of the maximum lead levels permitted under the Directive. The Committee was at the same time considering a motion tabled by Mr Johnson on 16 February 1982 which called on the Commission to propose amendments to present EEC rules to provide that all new cars put onto the Community market after 1 January 1985 be manufactured so as to take lead-free petrol and be required to run on such petrol; and to propose the necessary rules so that, as from 1 January 1985 a lead-free grade of petrol should be available at all places where petrol is sold.

3. CARBON MONOXIDE AND HYDROCARBONS

(a) Criteria and quality standards

Carbon monoxide in air was chosen as one of the pollutants for priority investigation on the grounds both of its toxicity and of the current state of knowledge of its significance in the health field.

Meetings of national experts and a European colloquium (17–19 December 1973) have been held to discuss and to critically analyse the available bibliography on the adverse or undesirable effects of the exposure of man to carbon monoxide. The results of this work are in general given in *Health Effects of Carbon Monoxide Environmental Pollution*. (Proceedings of the Colloquium.)

The Commission has also taken into account the work performed at national and international level. In particular it has considered the report of the WHO Expert Committee as it relates to carbon monoxide in the Technical Report Series No 506 *Air Quality Criteria and Guides for Urban Air Pollutants*.

A survey of the measured levels of carbon monoxide present in urban air of the Member States of the European Community has been made for the years 1971 and 1972. The results of this survey (rapporteurs P. Chovin and L. Truffert) are reported in the above-mentioned Proceedings of the colloquium.

(b) Reducing pollution at source

The principal source of pollution by carbon monoxide derives from the incomplete combustion of organic substances used as fuel, mainly due to vehicles powered by internal combustion engines (see Table 3.3).

Certain actions have already been taken on a Community basis to reduce air pollution by gases by positive ignition engines of motor vehicles. The first Council directive on this subject was adopted on 20 March 1970.

As far as the technical requirements were concerned, the Commission's proposal was based on those elaborated by the UN Economic Commission for Europe (ECE) in its regulation No 15 (Uniform provisions concerning the approval of vehicles equipped with a positive-ignition engine with regard to the emission of gaseous pollutants by the engine), annexed to the Agreement of 20 March 1958 concerning the adoption of uniform conditions of approval and reciprocal recognition of approval for motor vehicle equipment and parts. In the Commission's view, to use the work of ECE in this way could not be anything except an additional advantage for the industry of the (then 6) countries of the Community since a large number of states were represented in ECE.

The field of application of the Commission's proposed directive was as large as possible. It applied to any vehicle with a positive-ignition engine intended for use on the road, with or without body work, having at least four wheels, a permissible maximum weight of at least 400 kg and a maximum designed speed equal to or exceeding 50 km/h, with the exception of agricultural tractors and machinery and public works vehicles.

On 22 February 1974 the Commission submitted to the Council a proposal for a Council directive adapting to technical progress the Council directive of 20 March 1970. In the Commission's view, the application of the provisions of the Council directive of 20 March 1970 had already led to an important reduction of emissions of carbon monoxide and unburnt hydrocarbons from these vehicles within the Community. However, this effect had, in some measure, been offset by the constant growth in the

Table 3.3: Council directive of 14 July 1978
Air pollution by motor vehicles—revised carbon monoxide, hydrocarbon and nitrogen oxide limits

Reference weight	Mass of carbon monoxide per test in g	Mass of hydro-carbons per test in g	Mass of NO_x expressed in NO_2 equivalent per test in g
RW ≤ 750	78	7.8	10.2
750 < RW ≤ 850	85	8.2	10.2
850 < RW ≤ 1020	91	8.5	10.2
1020 < RW ≤ 1250	104	9.2	10.2
1250 < RW ≤ 1470	119	9.9	14.3
1470 < RW ≤ 1700	132	10.5	14.8
1700 < RW ≤ 1930	145	11.2	15.4
1930 < RW ≤ 2150	158	11.8	15.8
2150 < RW	172	12.5	16.3

vehicle population, with the result that in the Commission's view, it was necessary to increase the severity of the provisions of this directive. During this time, the findings of the ECE Working Group of experts 'Air Pollution due to motor vehicles – Technical Aspects' had shown that the intervening technical progress made in the construction of engines allowed for the further reduction, in the short term, of the admissible limits of emissions. Based on these findings, the Commission formulated a proposal for a directive modifying the above-mentioned directive. This proposal essentially provided for a 20% reduction in the limits relating to carbon monoxide, and a 15% reduction in those relating to unburnt hydrocarbons. In addition, the regulations concerning the test relating to emissions of carbon monoxide at idling speeds had been modified at all the possible settings of the carburettors. The Commission felt that together these modifications would, when applied, represent a new and important reduction in air pollution from motor vehicles and an important contribution to the improvement of the urban environment.

On 28 May 1974 the Council adopted a directive adapting to technical progress Council directive No 70/222/EEC of 20 March 1970.

Besides simplifying some of the provisions of the directive of 20 March 1970 so as to facilitate performance of the tests by the competent authorities and besides introducing certain amendments to the administrative procedure for the type-approval of a motor vehicle as regards the emission of pollutants, the new directive prescribed further reductions in the admissible level of emissions.

Two further adaptations to the 1970 directive have been adopted as Commission directives under the procedure for adaptation to technical progress. The first dated 30 November 1976 set limits for emissions of nitrogen oxides. The second, dated 14 July 1978, reduced limit values for all three pollutants. For details see Table 3.3.

Table 3.4: Commission proposal for Council Directive amending Council Directive of 20 March 1970 Air pollution by motor vehicles

Reference mass RW (kg)	Carbon monoxide L_1 (g/test)	Combined emission hydrocarbons and oxides of nitrogen L_2 (g/test)
RW ≤ 1020	58	19.0
1020 < RW ≤ 1250	67	20.5
1250 < RW ≤ 1470	76	22.0
1470 < RW ≤ 1700	84	23.5
1700 < RW ≤ 1930	93	25.0
1930 < RW ≤ 2150	101	26.5
2150 < RW	110	28.0

In April 1982 the Commission submitted to the Council a proposal for a Council directive amending the directive of 20 March 1970. The Commission took the view that a further reduction in limit values was required in order that the progress achieved through the implementation of earlier measures should not be cancelled out by increases in traffic density in highly urbanized areas of the Community. It found after examination in close cooperation with experts of national authorities and from the automobile industry, that the present state of technical development in motor vehicle construction would make it possible in a relatively short time to achieve, in relation to the limits laid down in the 1978 directive, a 23% reduction in CO emissions, and a 20 to 30% reduction, according to the weight category, in the combined HC and NO_x emissions. The proposal is based on these findings. A combined limit value for the permissible emissions of HC and NO_x is proposed because it gives motor vehicle manufacturers considerable leeway in the choice of engine modifications to be made for the purpose of reducing such emissions. Details of proposed limits are given in Table 3.4. The proposal also provides for:

1. the adoption of the constant volume sampling and analysis method at present used in the United States, Japan and Sweden;
2. including emissions from diesel engines in cars and light commercial vehicles.

The latter requires an amendment to the operative part of the directive of 20 March 1970 which cannot be made through the Committee on adaptation to technical progress. It must therefore be placed before the Council for adoption.

The proposals are based on the 04 series of amendments to Regulation

15 of the United Nations Economic Commission for Europe which were notified to signatory states in May 1981.

The Commission considers that any increase in petrol consumption of up to 5% resulting from these proposals could be offset and therefore not prevent manufacturers from fulfilling the voluntary undertakings to reduce consumption which they entered into at national level and which should result in a reduction of at least 10% in the fuel consumption of cars sold in the Community in 1985 compared with 1978.

4. SMOKE (Diesel Engines)

On 2 August 1972 the Council adopted a directive on the approximation of the laws of the Member States relating to the measures to be taken against the emission of pollutants from diesel engines for use in vehicles. This directive specifies that the emission of pollutants by the vehicle type submitted for approval shall be measured by the two methods described in the directive, relating respectively to tests at steady speeds and tests under free acceleration and that the emission of pollutants shall not exceed certain limits. These limits concern, amongst other things, the opacity of the exhaust gases produced by the engine.

On 8 December 1975 the Commission submitted to the Council a proposal for a similar Council directive on the approximation of the laws of the Member States relating to the measures to be taken against the emission of pollutants from diesel engines for use in wheeled agricultural or forestry tractors. This proposal was adopted by the Council on 28 June 1977.

5. NITROGEN OXIDES (NO$_x$)

(a) Preliminary Report of 3 April 1974

Nitrogen oxides were amongst the substances chosen for priority investigation under the first Action Programme on the Environment. In its preliminary report on the problem of pollution and nuisances originating from energy production the Commission gave a summary statement regarding the effects of nitrogen oxides on the environment. It was known that without the presence of nitrogen oxides in the atmosphere no photo-oxidation of hydrocarbons would occur and the development of photochemical oxidants would be much reduced. Sufficiently reduced levels of either NO$_x$, or hydrocarbons alone in the air tended to alleviate the formation of photochemical oxidants, but the exact relationships were extremely complex and not yet well understood.

Studies had shown that nitrogen dioxide might be associated with increased incidences of respiratory infections in children, with damage to vegetation and with corrosion of electronic components.

The report cited a review made by the Commission with a group of national experts, of the NO, NO_2 and NO_x ground level measurements carried out in 1971/1972 in Member States. Until then relatively few efforts had been made to measure these pollutants systematically. Furthermore, due to the variations in concentration with the exact siting and to the incompatibility of a number of analytical techniques, no comparison between the few results available could be made, and numerical indications of concentration ranges would not be significant.

Efforts were currently being made at the Community level to develop criteria for the siting of sampling stations and for the harmonization of the analytical methods.

In the few instances where trend analyses were available, upward trends seemed to be observed, showing the need to develop rapidly a harmonized set of stations within the European Community, generating concentration results upon which decisions could be made.

Nitrogen oxides were formed mainly during high-temperature combustion of fossil fuels in fixed as well as in mobile installations. Recent estimates indicated that the emission level would increase at about the same rate as fuel consumption, since there were no adequate control techniques now applied on a large scale for these compounds.

Emission factors (weight of NO, formed per unit weight of fuel consumed) varied considerably, not only with type of fuel and type of combustion unit, but also between apparently identical units burning the same fuel. Quoted emission factors could therefore only be average or typical values, and considerable variation was to be expected, depending on the mode of operation. This was particularly true of mobile sources, where engine load conditions were constantly changing.

A study had been completed under contract for the Commission's services aiming at an assessment of nitrogen oxide emissions from energy production within the EC Member States for the period 1970 until 1985, together with the possible abatement techniques and associated costs, taking into consideration existing studies and energy consumption forecasts.

Starting from weighted emission factors together with estimated fuel consumption in various sectors, it was concluded that the total NO_x emissions in 1970 within the EC Member States amounted to approximately 7.6 million tonnes. On the basis of the predicted fuel growth pattern used in the study, and assuming no controls, the total NO_x emissions were expected to rise to around 13 million tonnes in 1985 (see Table 3.5).

Analysis by sector of fuel use showed that in 1970 general industry accounted for one third of the emissions, but transportation was the fastest growing sector and was expected to take over one third of total NO_x emissions in 1985. Thus emissions from transport would double from 2.2 million tonnes in 1970 to 4.5 million tonnes in 1985, by which time it could replace industry as the largest source of nitrogen oxides.

The control technology of NO_x emissions was in its infancy, and much proving of techniques under real-life conditions remained to be done.

Table 3.5: NO_x emissions: Sector breakdown 1970 to 1985 (mean values)

Sector	1970		1980		1985	
	M tonnes NO_x	%	M Tonnes NO_x	%	M Tonnes NO_x	%
Electricity generation	2.1	28	3.1	28	3.5	27
Domestic	0.8	10	1.1	10	1.2	9
Industry	2.5	33	3.3	30	3.8	29
Transportation	2.2	29	3.6	32	4.5	35
TOTAL	7.6	100	11.1	100	13.0	100

(b) Council resolution of 3 March 1975 on energy and the environment

In its resolution of 3 March 1975 on energy and the environment, the Council took note of the Commission's preliminary report on the problems of pollution and nuisances relating to energy production and invited the Commission to submit proposals on policies to be followed by the Community and the Member States including, where nitrogen oxides were concerned, proposals for:

1. intensification of research relating to the effects of nitrogen oxides on man and the environment;
2. the development of methods for taking appropriate measures;
3. implementation of preventive measures to reduce sources of pollution by oxides of nitrogen pending advances in our knowledge of this field.

(c) Quality standards

The Commission is preparing a proposal for a Council directive setting quality standards for nitrogen oxides.

(d) Reducing pollution at source

Since 30 November 1976 nitrogen oxides emissions from motor vehicles have been covered by amendment to Council directive 70/222/EEC—see Section 3.b.

6. FLUOROCARBONS

As far as fluorocarbons in the environment are concerned, the Commission on 29 August 1977 submitted a draft Recommendation to the Council. This Recommendation was approved as a Resolution by the Council on 30 May 1978, and called upon the Member States:

● to pursue and intensify the actions already undertaken for a cooperation on a Community basis in planning research into the effects of
fluorocarbons on man and the environment and in making available
and interpreting the results;
● to take immediate steps to encourage all elements of the aerosol and
plastic-foam industries using chlorofluoromethanes F-11 and F-12 to
step up research into alternative products and to promote the development of alternative application devices;
● to take immediate steps to encourage industry and users of equipment
containing chlorofluoromethanes F-11 and F-12 to eliminate the leak-
age of these chemicals;
● to take all appropriate measures to ensure that there will no longer be
an increase in production capacity situated within the Community in
respect to the use of chlorofluoromethanes F-11 and F-12.

In the second half of 1978, the effects of fluorocarbons on the environment would be re-examined in the light of the available information, with
a view to reaching a Community policy.

At the International Conference on Chlorofluorocarbons which took
place in Munich 6–8 December 1978, Member States of the Community—
together with other participating countries—agreed to work for a significant
reduction in the release of CFCs in the next few years in relation to 1975
data. They also recognised that in the light of new and convincing scientific
evidence decisive reductions in the use of CFCs would be necessary. The
Federal Republic of Germany, host of the Munich Conference, reported to
the meeting of EEC Environment Ministers on 19 December 1978. The
Council invited the Commission to submit appropriate proposals as soon
as possible.

The Commission submitted to the Council a proposal for a decision on
chlorofluorocarbons in the environment on 16 May 1979. It was adopted
on 26 March 1980. It stipulated that the production capacity of CFCs 11
and 12 should not be increased in any Member State and that by 31
December 1981 there should be a reduction of at least 30% compared with
1976 levels of the use of the CFCs in aerosols.

Article 2 of that Decision provided for a re-examination during the first
half of 1980 of the scientific and economic data available. To this end, the
Commission forwarded a communication to the Council on 16 June 1980.

On 26 May the Commission sent to the Council a second Communication
concerning chlorofluorocarbons in the environment, containing information
and basis for evaluation for the pursuit of Community policy.

That Communication reviewed the available scientific and economic data
and concluded that the measures to be taken at Community level with
regard to CFCs should be as follows:

(a) Maintenance and consolidation of precautionary measures adopted
on 26 March 1980.

(b) Improvement in the collection of scientific, technical and economic data as a basis for a periodic review of Community policy.
(c) Engagement in projects, especially with the industry concerned, designed to decrease CFC emission in the sectors other than the filling of aerosol containers.
(d) Proposal of, and support for, suitable measures at international level.

The Ministers of the Environment meeting in the Council on 11 June 1981 debated and took note of this Communication and acknowledged the Commission's intention:

● to establish suitable procedures and mechanisms for an exchange of scientific, technical, socioeconomic and statistical information, and
● to undertake action in the sectors of foam plastics, refrigeration and solvents designed to reduce the emission of CFCs,

and recommended close collaboration between the Member States and the Commission on these points.

In addition, the Council instructed the Commission to submit at the earliest a proposal for a Council Decision on preventive and precautionary measures to be applied by the Member States after 31 December 1981, which should include:

● the reaffirmation of the zero growth of production capacity of CFC 11 and 12 on the basis of a precise definition;
● the consolidation of the reduction already achieved in the use of CFC 11 and 12 in the filling of aerosols;
● the possibility to proceed with a greater reduction in the aerosol sector;
● the preparation of measures for the reduction of CFC emissions in the sectors other than aerosols.

Finally, the Council invited the Commission to submit a proposal for a Council Decision authorizing the Commission to participate on behalf of the Community in the negotiations for a global framework Convention on the protection of the ozone layer.

The draft decision on preventive and precautionary measures in accordance with the Council's instructions (excluding the provision concerning negotiations for a Convention) was submitted to the Council on 7 October 1981. It was agreed by the Council on 24 June 1982.

On 19 January 1982 the Council adopted the decision authorizing the Commission to participate on behalf of the Community in negotiation for a global framework Convention on protection of the ozone layer.

7. FUTURE WORK

In adopting the Second Environment Action Programme of 13 June 1977, the Council called for a continuation of work on the determination of criteria, the fixing of quality objectives or standards, the exchange of information between surveillance and monitoring networks; and on measures relating to certain products, e.g. the exhaust gases of motor vehicles and the sulphur content of gas oils.

The Council also called on the Commission to study pollution problems arising in certain industrial sectors and in energy production, giving priority to industries emitting dust, oxides of sulphur and nitrogen, hydrocarbons and solvents, fluorine and heavy metals.

The Council decided that the Commission should also, through a panel of national experts, organize exchanges of information on ways of combating atmospheric pollution at national or regional level, particularly:

● implementation of national or regional plans;
● establishment of administrative and scientific bodies responsible for air management;
● use of economic measures;
● listing of sources of pollution;
● organization of a procedure for the exchange of information betweeen early warning networks;
● use of mathematical models;
● establishment of standards for certain pollutants regarded as dangerous;
● monitoring of establishments causing pollution.

The Commission would also examine problems relating to:

● the influence of fluorine and chlorine compounds and nitrogen oxides on the upper layers of the stratosphere;
● pollution resulting from the increasing use of gas turbines;
● pollution problems arising from the use of small installations (domestic heating equipment, incineration appliances, etc.);
● the possible effects of transfrontier pollution.

In the light of the results of these exchanges of information and studies, and on the basis of work done by other international organizations, the Commission would, if necessary, submit appropriate proposals to the Council.

In its proposal for a draft Action Programme for 1982–1986, the Commission proposes:

● further efforts to establish air quality standards to control atmospheric pollution;

- further study of the possibility of drawing up Community standards for pollutants which are produced by a large number of scattered sources i.e. nitrogen oxides and hydrocarbons;
- study of the effectiveness of regionally applied standards for pollutants produced by a limited number of sources with a small radius of effect such as fluorine, cadmium, mercury and carbon monoxide;
- for some ubiquitous pollutants, devising of a policy initially to stabilize and thereafter gradually reduce total emissions by establishing emission standards where necessary, for certain sources, notably large fixed sources with high stacks;
- specific measures to reduce the discharges of pollutants from burning coal;
- study of how to strengthen measures to control pollution from motor vehicles;
- continuation of work on harmonization of measuring methods;
- study of the effects of certain chemicals such as chlorofluorocarbons on the ozone layer in the stratosphere and on the climate.

The Commission put forward in a document dated 11 June 1981 (COM(81)317 Final) dealing with the European automobile industry a general approach for the formulation of Community rules affecting motor vehicles, based on a detailed examination of the economic, environmental and social effects of the measures envisaged in the field of pollution emissions. The necessary machinery was set up in January 1982 by the creation of the Working Party on Air Pollution. This Working Party has 18 months in which to examine the technical possibilities for reducing pollution emissions and to assess the repercussions thereof, notably with regard to fuel consumption and vehicle manufacturing and servicing costs.

4

Waste

Under the section 'Industrial and Consumer Wastes' of Chapter 7 'Action concerning wastes and residues' of Title 1 of Part 2 of the Environment Programme, the Council stated that the most important problem for the Community in this field was the elimination of wastes which, because of their toxicity, their non-degradability, their bulk, or for other reasons, require a solution extending beyond the regional framework and possibly even beyond national framework. Even if the harmful effects of the wastes do not extend beyond the immediate region, Community action may well become necessary if the elimination or re-use of the wastes is dependent on economic resources. If the solutions adopted give rise to differences in the production and distribution conditions of certain goods, these differences may have repercussions on the functioning of the common market and on international trade.

The Environment Programme specified that work to be carried out should cover: the drawing up of a qualitative and quantitative inventory of wastes or residues which are particularly harmful to the environment; a study of the economic and legal aspects of the problems posed by the collection, transport, storage, recycling or final treatment of particular wastes, including toxic and dangerous waste, waste oils, titanium dioxide wastes, bulky ferrous scrap, non-biodegradable packaging and waste from slaughter houses and breeding establishments; and, an examination of the action to be taken at Community level with regard to these wastes. The Council called upon the Commission to make appropriate proposals.

1. FRAMEWORK DIRECTIVE

On 10 September 1974 the Commission submitted to the Council a proposal for a Council directive on waste disposal. The directive, as adopted by the Council on 15 July 1975, defines 'waste' as any substance or object which the holder disposes of or is required to dispose of pursuant to the provisions of national law in force. 'Disposal' is defined as the collection, sorting, transport and treatment of waste as well as its storage and tipping above or underground, the transformation operations necessary for its re-use, recovery or recycling.

The fundamental obligation of the directive is that Member States are required to take the necessary measures to ensure that waste is disposed of without endangering human health and without harming the environment, and in particular:

- without risk to water, air, soil and plants and animals;
- without causing a nuisance through noise or odours;
- without adversely affecting the countryside and places of special interest.

Member States are required to establish or designate the competent authority or authorities to be responsible in a given zone for the planning, organization, authorization and supervision of waste disposal operations.

These competent authorities are required to draw up as soon as possible one or several plans relating to, in particular:

- the type and quantity of waste to be disposed of;
- general technical requirements;
- suitable disposal sites;
- any special arrangements for particular wastes.

The plan or plans may, for example, cover:

- the natural or legal persons empowered to carry out the disposal of waste;
- the estimated costs of the disposal operation;
- the appropriate measures to encourage rationalization of the collection, sorting and treatment of waste.

Member States are required to take the necessary steps to ensure that any holder of waste:

- has it handled by a private or public waste collector or by a disposal undertaking;
- or disposes of it himself in an ecologically inoffensive manner.

The directive provides that any installation or undertaking treating, storing or tipping waste on behalf of third parties, must obtain a permit from the competent authority, relating in particular to:

- the type and quantity of waste to be treated;
- general technical requirements;
- precautions to be taken;
- the information to be made available at the request of the competent

authority concerning the origin, destination and treatment of waste and the type and quantity of such waste.

The installations and undertakings referred to are to be periodically inspected by the competent authority to ensure, in particular, that the conditions of the permit are being fulfilled.

Member States are required every three years to draw up a situation report on waste disposal in their respective countries and forward it to the Commission. The directive imposes an obligation on the installations or undertakings in disposing of waste to supply the competent authority with the necessary information. The Commission is to circulate this report to the other Member States and to report every three years to the Council and to the European Parliament on the application of the directive.

The directive also contained a provision that Member States should take appropriate steps to encourage the prevention, recycling and processing of waste, the extraction of raw materials and possibly of energy therefrom and any other process for the re-use of waste.

Member States are also to inform the Commission in good time of any draft rule concerning:

(a) the use of products which might be a source of technical difficulties as regards disposal or lead to excessive disposal costs;
(b) the encouragement of:
   ● the reduction in the quantities of certain waste;
   ● the treatment of waste for its recycling and re-use;
   ● the recovery of raw materials and/or the production of energy from certain waste;
(c) the use of certain natural resources, including energy resources, in applications where they may be replaced by recovered materials.

2. WASTE MANAGEMENT COMMITTEE

By its decision of 21 April 1976, the Commission established a Committee on Waste Management (CWMC). The task of this Committee is to supply the Commission with opinions, either at the request of the Commission or on its own initiative on all matters relating to:

(a) the formulation of a policy for waste management having regard to the need to ensure the best use of resources and the safe and effective disposal of waste;
(b) the different technical, economic, administrative and legal measures which could prevent the production of wastes or ensure their re-use, recycling or disposal;
(c) the implementation of directives on waste management and the formulation of fresh proposals for directives in this field.

The Committee is chaired by a representative of the Commission and consists of 20 members, two from the Commission and two from each Member State.

The Committee met for the first time in March 1977. It listed the following items for priority consideration: toxic wastes, waste paper, packaging, utilization of waste as a fuel and its economic use in agriculture. The secondary priorities listed by the Committee are agricultural wastes, demolition waste and textile waste. At its second meeting on 4 October 1977 the CWMC approved action programmes on packaging and waste paper. It also decided on the terms of reference of the working parties to be set up to investigate the energy potential of wastes and their use in agriculture.

At a meeting on 9 and 10 March 1982 the Committee agreed on new directions to take. These covered the transfrontier transfer of toxic wastes, criteria for disposal of waste, burning of waste oil and the setting up of a data bank on waste.

3. WASTE OILS

Waste oil and residues containing petroleum and tar, in particular residues containing lubricants, are, as noted, among the substances listed as deserving priority attention under Chapter 7 'Action concerning wastes and residues' of the Environment Programme. A study of this problem, carried out for the Commission, came to the conclusion that the pollution of soil and water by waste oils was becoming acute due to growing industrialization, urbanization, and the continued development of transport facilities. In addition, certain treatments of waste oils created new sources of pollution, especially air pollution.

There had been a steady increase in the quantity of waste oils and in particular of emulsions, a large part of which were disposed of without controls.

The extent and urgency of the problem was underlined by the fact that sometimes as much as 20–60% of all waste oils were disposed of without any control in some Member States; the resulting water pollution would account for approximately 20% of all industrial pollution according to some estimates.

The percentage of recovery was therefore about 50%; according to information received, only about 1 million tonnes of waste oils was actually recycled, thus 1 million tonnes was lost as energy or as lubricants, with obvious consequences for the environment and for a comprehensive fuel supply policy.

In view of the results of this study, and in view of the fact that the French and Dutch Governments had sent to the Commission, under the Information Agreement of 5 March 1973, legislative proposals relating, inter alia, to the disposal of waste oils, the Commission prepared and submitted to the Council a proposal for a directive on the disposal of waste oils.

The Commission's draft directive, based on Article 100 of the Treaty of the EEC, was intended to harmonize legislation and to thus create a coherent system of legal provisions applicable in all Member States. It also had the double objective of ensuring environmental protection and at the same time the maximum possible re-use of waste oils which, in the Commission's view, could make a large contribution to energy supply policy.

The directive on the disposal of waste oils, as adopted by the Council on 16 June 1975, defines 'waste oils' as any semi-liquid or liquid used product totally or partially consisting of mineral or synthetic oil, including the oily residues from tanks, oil-water mixtures and emulsions.

Member States are required to take the necessary measures to ensure the safe collection and disposal of waste oils and to ensure that, as far as possible, the disposal of waste oils is carried out by recycling (regeneration and/or combustion other than for destruction).

If the fundamental objectives of the directive cannot otherwise be attained, Member States are required to take the necessary measures to ensure that collection and disposal operations are in fact carried out. They are empowered under the directive to grant indemnities to collection and/ or disposal undertakings for the service rendered. These indemnities may be financed, among other methods, by a charge imposed on products which after use are transformed into waste oils, or are waste oils.

4. POLYCHLORINATED BIPHENYLS (PCBs)

The Action Programme spoke of the important problem posed for the Community by the elimination of wastes which, because of their toxicity, their non-degradability, their bulk or for other reasons, required a solution extending beyond the regional framework and possibly even beyond national frontiers.

On 10 February 1975 the Commission submitted to the Council a proposal for a Council directive on the collection, regeneration and/or destruction of polychlorinated biphenyls (PCBs). The Council adopted this directive on 6 April 1976.

The aim of the proposed directive was to cover the conditions of collection, regeneration and destruction of PCB and thus to supplement the control of these substances in order to avoid any dispersal in the environment. The Commission noted that there were at the time no specific laws on collection, regeneration and/or destruction of PCB in force in the Member States.

Under the directive Member States are required to take the necessary measures to prohibit the uncontrolled discharge, dumping and tipping of PCB and of objects and equipment containing such substances, and also to make compulsory the disposal of waste PCB and PCB contained in objects and equipment no longer capable of being used.

The preferred method of disposal is regeneration.

Member States are required to set up or designate the installations, establishments or undertakings which are authorized for the purposes of disposing of PCB on their own account and/or on behalf of third parties.

Anyone holding PCB who is not authorized to dispose of it himself is required to hold it available for disposal by the authorized installations, establishments or undertakings.

As is also the case with the waste oil directive, the directive establishes that the cost of disposing of PCB shall be borne in accordance with the 'polluter pays' principle.

Member States are required to draw up every three years a situation report on the disposal of PCBs in their territory within the framework of the general report on the disposal of waste which is provided for in the framework directive on waste adopted by the Council on 15 July 1975.

5. TOXIC AND DANGEROUS WASTES

The Council, in adopting the European Communities Programme of Action on the Environment, acknowledged that the disposal of toxic and dangerous wastes was one of the most important problems for the Community and that it therefore required 'a solution extending beyond the regional framework and possibly beyond national frontiers'.

In view of the specific nature of this problem, the Council therefore decided that the Commission should study its technical, economic and legal aspects, examine the action to be taken at Community level and submit to the Council conclusions from this work together with the proposals arising out of it.

Assisted by a Working Group of National Experts and a Sub-Group of Scientific Experts on toxic and dangerous wastes, and with the aid of certain studies commissioned specifically for this purpose, the Commission reviewed some of the major problems arising in the disposal of toxic and dangerous wastes.

On 22 July 1976, the Commission submitted to the Council a draft proposal for a Council directive on toxic and dangerous wastes. The directive was adopted by the Council on 12 December 1977. The directive defines a common field of action within which rules on toxic and dangerous waste disposal are to be applied. The concepts of 'toxic and dangerous waste' and of 'disposable' are defined and obligations are imposed on Member States to ensure that disposal operations are carried out without endangering human health and the environment.

Member States are furthermore required to take the necessary measures to encourage the recycling and processing of toxic waste, the extraction of raw materials and possibly of energy therefrom.

The toxic and dangerous substances or materials specifically envisaged in the directive are given in Table 4.1.

Table 4.1: Council Directive of 20 March 1978
on Toxic and Dangerous Wastes
List of toxic or dangerous substances and materials

This list consists of certain toxic or dangerous substances and materials selected as requiring priority consideration
1 Arsenic: arsenic compounds
2 Mercury: mercury compounds
3 Cadmium: cadmium compounds
4 Thallium: thallium compounds
5 Beryllium: beryllium compounds
6 Chrome 6 compounds
7 Lead: lead compounds
8 Antimony: antimony compounds
9 Phenols: phenol compounds
10 Cyanides, organic and inorganic
11 Isocyanates
12 Organic-halogen compounds, excluding inert polymeric materials and other substances referred to in this list or covered by other Directives concerning the disposal of toxic or dangerous waste
13 Chlorinated solvents
14 Organic solvents
15 Biocides and phyto-pharmaceutical substances
16 Tarry materials from refining and tar residues from distilling
17 Pharmaceutical compounds
18 Peroxides, chlorates, perchlorates and azides
19 Ethers
20 Chemical laboratory materials, not identifiable and/or new, whose effects on the environment are not known
21 Asbestos (dust and fibres)
22 Selenium: selenium compounds
23 Tellurium: tellurium compounds
24 Aromatic polycyclic compounds (with carcinogenic effects)
25 Metal carbonyls
26 Soluble copper compounds
27 Acids and/or basic substances used in the surface treatment and finishing of metals

The directive lays down that, with certain exceptions, wastes containing the above toxic and dangerous substances or materials can be disposed of only by the installations, establishments or undertakings authorized by the competent national authorities to do so on their own account or on behalf of third parties.

It also establishes that any holder of toxic waste, who has not been granted such an authorization, is required to deliver the waste, to an authorized installation.

The directive does not lay down the specific methods of disposal for the various categories of toxic and dangerous waste. However, the Commission's role in implementing the directive will include, subsequently, the elaboration of codes of practice for the disposal of various toxic and dangerous wastes.

In order to ensure maximum coordination at national and Community level, it is foreseen that special plans for the disposal of toxic and dangerous waste shall be drawn up and kept up to date by the competent national authorities. Member States shall forward them to the Commission and draw up every three years a situation report on the disposal of toxic and dangerous waste in their respective countries.

The Commission will itself report every three years to the Council and to the European Parliament on the implementation of the directive.

As is the case with other Directives adopted in the field of environment, for example those relating to the quality of water for bathing or for freshwater fish, or to the exhaust gases of motor vehicles, the Council Directive on toxic and dangerous wastes provides for a mechanism for adapting the Directive to technical and scientific process. The procedure laid down is that established under the Council Resolution of 15 July 1975, (see Annex III) which provides for the creation of a Technical Adaptation Committee empowered to take decisions on the basis of a qualified majority. The Directive specified that the amendments necessary for adapting this Directive to scientific and technical progress should be:

- to state the name and composition of the toxic and dangerous substances and materials covered by the Directive.
- to add toxic and dangerous substances and materials unknown at the time of notification of this Directive.

In adapting the Directive to technical and scientific progress, account should be taken of the immediate or long term hazard to man and the environment presented by waste by reason of its toxicity, persistence, bioaccumulative characteristics, physical and chemical structure and/or quantity.

When the Council adopted the Directive, the Commission stated that the amendment procedure was intended to be applied only in respect of adaptations of a scientific or technical character. Referring to point B of the Council Resolution of 15 July 1975, the Commission stated moreover that it would not initiate the procedure in respect of adaptations which might raise major issues of an economic or political nature for one or more Member States and that in this case it would submit appropriate proposals direct to the Council. In deciding whether such issues arose, the Commission would take account of any views expressed, in the course of prior consultation, by members of the Technical Adaptation Committee.

The Commission pointed out that this statement in no way affected the normal rules governing the operation of the Technical Adaptation Committee and that it could not be interpreted as offering Member States the possibility of opposing referral by the Commission to the Committee of the draft decisions concerned.

6. TRANSFRONTIER SHIPMENT OF HAZARDOUS WASTES

On 10 January 1983, the Commission sent to the Council a proposal for a Directive on the supervision and control of the transfrontier shipment of hazardous wastes within the Community. The Commission noted that for several years now transfrontier shipments of hazardous wastes had been increasing steadily, primarily for the following reasons:

- because of the lack of suitable firms, plants and dumps or of sufficient capacity in the country in which the wastes originated;
- because plants or dumps in the country of destination were closer at hand than those in the country in which the wastes originated;
- because disposal costs, including transport costs, were lower in the country of destination, since more advanced technologies were available or less stringent environmental legislation applied there;
- to evade more stringent national controls and requirements.

The proposal for a Directive on the transfrontier shipment of hazardous wastes was intended to supplement the Directives on waste already adopted by ensuring uninterrupted supervision and monitoring of hazardous wastes from the source right through to their final non-polluting disposal, even if this was to take place on the other side of the national frontier, in the interests of protecting human health and the environment.

Accordingly, it would be mandatory for the consignor to notify the competent authorities in the country of dispatch, in the destination country and in the country of transit, if any, whenever hazardous wastes were to be shipped to another Community country.

Article 14(2) and (3) of the Council Directive 78/319/EEC on toxic and dangerous waste, which stipulated that these wastes must be accompanied by an identification form during transport, was to be extended explicitly to include transfrontier shipments. As a result, the wastes covered by this Directive (see Table 4.1) would have to be accompanied by a standard Community consignment note each time they were shipped across national frontiers.

Waste oils were also included in the Commission's proposed Directive, since recent research into the carcinogenic effects of used engine lubricants and coolants had confirmed that they were toxic and dangerous.

The Commission noted that the Waste Management Committee, which the Commission set up in 1976 to advise it on matters relating to the formulation and implementation of the Community's waste management policy, had approved the broad lines of the proposal for a Directive and at the same time had stressed that it was urgent for the Council to approve the proposal.

Within the Community only the Federal Republic of Germany had specific legislation on the transfrontier shipment of hazardous wastes in the form of the Order of 29 July 1974 concerning the import of wastes (i.e. the Abfalleinfuhr-Verordnung). None of the other Community countries had specific legislation concerning waste imports or exports. All that existed was a series of *ad hoc* bilateral agreements between the competent authorities in individual Community countries, in an attempt to guarantee a certain degree of supervision of transfrontier shipments of hazardous wastes.

Having held extensive meetings with government experts and specialists from industry, the Commission stated that all the Community Member States felt that these bilateral agreements were not fully satisfactory and that uniform legislation on the transfrontier shipment of hazardous wastes within the Community was urgently needed. Some 160 million tonnes of industrial waste were produced in the European Community each year; between 25 million and 30 million tonnes came under the heading of toxic or dangerous wastes. According to the available statistics, roughly 10% of these hazardous wastes were transported across national frontiers before disposal or treatment.

At present supervision of these wastes ends at the national frontiers, even in those countries which already had an extensive monitoring system. The countries to which the wastes were shipped generally knew nothing about the hazardous wastes imported into their territory nor did they have sufficient control over them, since the wastes were frequently inadequately specified and identified and, very often, even classified differently. This has resulted in a series of accidents, leading to the gradual realization that transfrontier shipment of hazardous wastes presented an exceptionally high risk to human health and to the environment.

7. SEWAGE SLUDGE

Interest in the subject of sewage sludge goes back to the earliest days of the Community Environment Programme. A specific research project was set up under the COST Programme (involving Community and other countries with a Commission secretariat) to study sewage sludge. Initially it was to cover the period 1973–74, and a second three year programme to follow on was adopted on 27 September 1977. Subsequently it was absorbed as a project under the 1980–84 Programme of Environmental Research.

Both first and second programmes deal in a general way with waste management but there is no specific reference under these heads to sewage sludge. In the second programme (OJ C139, p. 22 13 June 1977) there is reference under a section dealing with intensive stock rearing to the Commission studying health requirements and permissible maximum levels for undesirable substances in stock raising wastes intended for spreading, as well as those relating to other forms of organic wastes used for the same purpose.

In its report on 'Progress made in Connection with the Environment Programme' (COM (80) 222) the Commission said a proposal governing the spreading of sludge was being prepared which would concern soil enrichment and protection (classification problems caused by heavy metal content and the purpose for which sludge is used.)

The Waste Management Committee set up by the Commission in 1976 considered the use of waste in agriculture one of the priority areas for action at Community level. A working party on the Use of Waste in Agriculture composed of Government experts was set up to advise the Commission on Community measures aimed at promoting the use of urban and similar refuse for agricultural purposes and eliminating nuisances and pollution from spreading waste on crop growing land.

On 10 September 1982, the Commission sent a proposal for a Council Directive on the use of sewage sludge in agriculture which would promote the use of certain types of solid waste in agriculture by stressing their agronomic value and would specify the precautions necessary for their proper use in agriculture to avoid pollution. It started from the position that in a number of Community countries guidelines existed (with or without legal backing) for the spreading of sewage sludge (Germany, UK, Denmark). France had a provisional standard and approval procedures, the Netherlands was drafting legislation while the remainder had no legislation or specific provisions.

In 1979 it was estimated that the annual amount of raw sewage sludge was about 230×10^6 tonnes—some 13% of the total waste stream—but expressed as dry matter the amount was 6×10^6 tonnes. With the substantial increase in sludge from treatment plants, estimates for 1990 could be as much as 15 to 20×10^6 tonnes of dry matter per year.

At that date 45% was dumped in landfill sites, 7% was incinerated, 19% was dumped at sea, 29% was used in agriculture. The figure of use for agriculture was felt to be low.

Possible land application depended on the treatment the sludge had undergone. Nitrogen and phosphorus were usually the nutrients most abundantly available in sludge and it was concluded that their levels could be used to establish the rate at which sludge might be spread, but the actual amount applied would depend on the level of trace elements.

Adverse effects from excessive addition of trace elements to soil might arise but the toxicity of these elements (particularly heavy metals) depended primarily on the type of soil and the receiving crops.

The proposal would set mandatory and/or recommended values for a number of trace elements with regard to (a) the concentrations of those elements permissible in the sludge itself; (b) the total quantity of those elements which may be added over a ten-year period; and (c) the concentrations admissible in soil to which sludge is applied. It would also strictly limit the use of unstabilized sewage sludge; prohibit spreading for certain types of crops; specify a minimum period between spreading and grazing or harvesting; impose limits on the pH value of treated soils; require special

authorization for applications to parks, playgrounds and woodlands; require regular analyses (along specified lines) of sludge and of treated soils;
and require records to be kept of sludge supplied for agricultural use. The
draft directive would, however, only apply to sludge from sewage treatment
works dealing with (a) urban effluents from a population of above 5,000,
or (b) mixed urban and industrial effluents (see Table 4.2)

Table 4.2
Annex 1A: Limit values (Recommended or Mandatory) of trace-element concentrations in sludge for use in agriculture (mg/kg dry matter)

ELEMENT		R	M
Cadmium	(Cd)	20	40
Copper	(Cu)	1000	1500
Nickel	(Ni)	300	400
Lead	(Pb)	750	1000
Zinc	(Zn)	2500	3000
Chromium	(Cr)	750	—
Mercury	(Hg)	16	—

Annex IB: Amounts of trace elements which may be added annually to agricultural land, based on a ten-year average (kg/ha/yr)

ELEMENT		R	M
Cadmium	(Cd)	0.10	0.15
Copper	(Cu)	10	12
Nickel	(Ni)	2	3
Lead	(Pb)	10	15
Zinc	(Zn)	25	30
Arsenic	(As)	0.35	—
Chromium	(Cr)	10	—
Mercury	(Hg)	0.40	—

Annex IC: Concentrations of trace elements admissible in agricultural soil to which sludge is applied (mg/kg dry matter)

ELEMENT		R	M
Cadmium	(Cd)	1	3
Copper	(Cu)	50	100
Nickel	(Ni)	30	50
Lead	(Pb)	50	100
Zinc	(Zn)	150	300
Arsenic	(As)	20	—
Chromium	(Cr)	50	—
Mercury	(Hg)	2	—

Annex IIA: Sludge analysis

As a rule, sludge is to be analysed at least every six months. Where changes occur in the characteristics of the sewage being treated the frequency of the analyses must be increased. If the results of the analysis do not vary significantly over a full year, the sludge must be analysed every 12 months. Analysis should cover the following parameters:

- dry matter, organic matter
- pH
- nitrogen and phosphorus
- copper, cadmium, nickel, lead and zinc.

It is for the Member States to decide whether the following parameters should also be calculated:

- C/N ratio
- salinity
- faecal content.

Analyses must also be carried out for the other trace elements for which the Member States have fixed recommended values.

Annex IIB: Soil analysis

Soils must be analysed prior to the first application of sludge and every five years thereafter.*

Analysis should cover the following parameters:

- pH
- copper, cadmium, nickel, lead and zinc

It is for the Member States to decide whether the following parameters should also be calculated:

- physical characteristics of the soil
- buffering capacity.

Analyses must also be carried out for the other trace elements for which the Member States have fixed recommended values.

* The number of soil samples taken will depend on the homogeneity of the soil and the land area; the depth at which samples are taken will depend on the type of crop.

8. WASTE PAPER

On 14 May 1980 the Commission submitted to the Council a draft recommendation concerning the recovery and use of waste paper and board. Three main aspects to the problem of re-using waste paper and board were recognised in the proposal. Firstly it aimed to carry out the tasks set out in the Action Programme on the Environment. It was secondly a logical extension of the action begun by the directive on waste of 15 July 1975 (75/442/EEC) in that it gave precision to the general policy aims set out in that framework directive. Finally it recognized that there were important and economic problems to be solved. The recommendation was adopted on 3 December 1981. In it the Council recommends member states and the Community Institutions to promote the use of recycled and recyclable paper and board.

9. BEVERAGE CONTAINERS

On 23 April 1981 the Commission submitted to the Council a draft directive on containers of liquids for human consumption. The purpose of the proposed Directive is to reduce the environmental impact of used containers and to reduce the consumption of energy and raw materials in this field. Its aim is to reduce the amount of containers of liquids for human consumption in household waste and to encourage a better recovery thereof.

In the Commission's view, it was quite clear from all the evidence that an operation of this kind must be carried out in an harmonized way throughout the Community. Individual operations carried out in a random way by the Member States could only lead to unfavourable consequences on the free movement of goods.

The proposal would require Member States, in respect of their own situation, to fix objectives for the purpose of reducing the amount of waste, to progressively attain these objectives, to use a number of means to achieve this end, and to inform the Commission of the objectives adopted, the means used and the results obtained.

The choice of means to be used to attain the goal which is fixed would be left to the Member States. They may opt for voluntary agreements or for legislation. They may encourage the recycling of used containers and/or their refilling for further use. It would therefore be more a question of defining a context within which Member States are to act than of laying down specific and detailed provisions.

The proposal would cover most liquids sold in containers for human consumption e.g. wine, beer, spirits, milk, edible oils, non-alcoholic beverages and fruit juices. 'Containers' are defined as any bottle, can, jar, carton or other type of sealed container containing liquid for human consumption.

10. RADIOACTIVE WASTE: A SPECIAL CASE

The Commission's constant interest in the radioactive waste problem is reflected in specific steps taken over a number of years. As early as 1965, it granted considerable financial aid towards the construction of a vault for experimental waste storage in the salt mine at Asse.

But it was not until 1973 that the major Community programmes were launched within the framework of the first Environment Programme. These have been implemented in the form of 'direct action', i.e. directly by the Community's Joint Research Centre (JRC), whose work is entirely financed by the Commission, or in the form of 'indirect action', i.e. through a series of shared-cost contracts between the Commission and public or private bodies in the Member countries.

The first multiannual programme of direct action on radioactive waste, costing 6.9 million u.a.,* was begun in 1973 and completed in December 1976. Most of the studies were carried out at the Ispra establishment, being concerned in particular with the determination of the long-term hazards presented by wastes and the separation and transmutation of actinides. The second programme, costing 21.06 million u.a., was undertaken as part of the JRC's multiannual programme for 1977–1980. The work represents an extension of the studies begun earlier, with emphasis on the assessment of long-term risks. Work continues under the 1980–1983 JRC programme.

In June 1975, the Council of Ministers of the Community adopted the first multiannual programme of indirect action, providing for a Community contribution of 19.6 million u.a., for the period 1975–1979. This contribution represents about 40% of the total cost of the programme, amounting to approximately 50 million u.a. Work will be principally concerned with the treatment of radioactive waste, disposal of such waste in geological formations and the solution of administrative, legal and financial problems of waste management.

The management structure required for these two programmes had to be flexible to take account of the widely differing characteristics of the participants and unified to ensure efficiency. The programmes are therefore implemented under the Commission's responsibility with the assistance of an Advisory Committee on Programme Management (ACPM), which serves both programmes and its standing working parties, comprising national officials with direct responsibility for research, follow the progress of the work and discuss it with Commission representatives so as to ensure a supply of up-to-date information to the laboratories and effective coordination.

These programmes are ambitious but not unreasonably so. They form a coherent entity, backed by substantial funds and involving research facilities

* In 1973 the unit of account (u.a.) was defined as the value of 0.88867088 grams of fine gold using conversion rates corresponding to the parities declared to the International Monetary Fund.

and personnel both in the Member States and in the competent departments of the Commission.

The direct action programme comprises desk studies and experimental work. Emphasis is placed on three topics.

(a) chemical separation and nuclear transmutation, in order to gain a better understanding of these new radioactive waste management techniques based on the recycle of elements presenting long-term hazards;

(b) evaluation of the long-term hazards of radioactive waste storage, in order to assess and define the long-term safety of ultimate disposal in suitable geological formations;

(c) the decontamination of reactor components.

The first indirect action programme covered:

(a) Research aimed at solving technological problems involved in the treatment, storage and disposal of radioactive waste.

(b) Measures to help in defining general arrangements (legal, administrative and financial) for waste storage and disposal operations.

The Commission is also studying the problems raised by the decommissioning of power reactors. This study is being carried out with the assistance of a group of national experts under the Second Programme of Action on the Environment, adopted by the Council in December 1976. It should serve as a basis for mapping out a programme of indirect R & D action and laying down guiding principles to govern decommissioning.

On 24 August 1977, the Commission sent to the Council a Communication on a Community plan of action in the field of radioactive wastes. The Commission noted that so far Community action had basically consisted of the research and development programmes currently in progress in the Community. The Commission was of the view that these programmes represented an initial step which must nevertheless be followed by others, with a suitable back-up, if growing volumes of radioactive waste were to be satisfactorily handled by 1990 and 2000.

On 5 March 1979 the Commission submitted to the Council a proposal for a decision adopting a second five year programme on radioactive waste management and storage to commence on 1 January 1980. The proposed programme consisted of four sections: treatment and conditioning of radioactive waste; storage and disposal of radioactive waste; evaluation of processes, tentative criteria and waste management strategies; and studies on the legal, administrative and financial aspects of waste management. The proposal was adopted on 13 March 1980.

11. FUTURE WORK

In adopting the Second Environment Action Programme, the Council stated that the protection of the environment against pollution, sound economic management of resources, the effort to reduce the Community's dependence on imported raw materials, the rational long-term management of natural resources which are either non-renewable or can be renewed only at a certain rate—all these considerations together argue in favour of an immediate and hard-hitting campaign against waste.

This campaign must be directed:

- at the consumer, who by his attitude and behaviour plays an important role in the generation of waste;
- at industry, which is anxious to reclaim costly raw materials used in the production processes and interested in the possibility of developing new recovery systems, but at the same time, by reason of the volume and range of its products, is responsible for a large proportion of the waste generated. Industry should also plan measures to extend the durability of its products in order to reduce the generation of waste;
- at local authorities, which are responsible for the collection and, where appropriate, the sorting of waste;
- at the national authorities, which can, for example by means of public procurement contracts, play an important role in increasing outlets for certain reclaimed substances and, more generally, can introduce overall policies for the rational use of raw materials.

The Council noted that all the Member States were aware of the need for action to improve the recovery of materials contained in waste.

In all Member States bodies had been set up to study such recovery and to define priorities. In some cases specialized bodies had been established to carry out a number of specific projects.

The Community was intimately concerned with these problems and ought for many reasons to promote an active anti-waste policy:

(a) to reduce pollution arising from the unorganized accumulation and unsuitable processing of waste;

(b) to contribute to the harmonious development of the economic activities entrusted to it by the EEC Treaty; such development cannot avoid the negative impact of the increase in the cost of raw materials, of the dependence of the Community and its Member States on external sources of supply and, in the long run, of the foreseeable depletion and resulting predictable rise in cost of certain materials;

(c) to avoid distortions of competition and obstacles to trade which might arise from measures taken to deal with waste solely at national level (for example, aid, dues and taxes, transfer of toxic waste from one State to another, or prohibition of such transfer);

(d) to disseminate knowledge of the problems concerned and of action taken at the various decision-making levels, and so to put into effect as efficiently as possible and at the appropriate levels the most suitable legal, technical and economic solutions.

The Council noted that the Community had already taken a number of decisions on the recovery and disposal of waste and that the Commission would continue to put into effect all these Council decisions. More generally, the Community's waste programme would include measures to encourage and improve waste recycling and re-use operations and the study and adoption of measures to prevent the generation of waste and to ensure the disposal of non-recoverable residual waste in ways which do not hold dangers for man or the environment.

In its proposals for the Third Environment Action Programme 1982 the Commission proposes that the Community should continue its action on waste management described in the Resolution of 17 May 1977 (under which the Second Action Programme was adopted), bearing in mind the following three main objectives of waste management policy in all areas of activity:

● to reduce the quantity of non-recoverable waste, and ultimately to abolish it;
● to recover, recycle and re-use waste for raw materials and energy;
● to manage non-recoverable waste properly and dispose of it in a harmless manner.

Greater emphasis should be placed on the recovery, recycling and re-use of waste, and on the prevention of waste and on product design which facilitates recycling.

In accordance with the guidelines laid down by the Committee on Waste Management; this action should concentrate mainly on the agricultural and energy uses of waste.

As far as waste disposal is concerned, Community rules on the management of toxic and dangerous wastes in particular should be supplemented and reinforced, by progressively substituting re-use for disposal of this waste.

As to the prevention of waste, the development of new technology which facilitates waste recycling or which removes the production of waste should be encouraged.

5
Chemicals in the Environment

1. EXISTING CHEMICALS

In the Programme of Action of the European Communities on the Environment it was recognized that regulations governing the classification, packaging and labelling of dangerous substances and preparations were not necessarily sufficient in all cases; it might be necessary also to prohibit or restrict marketing and use under certain conditions.

On 25 July 1975 the Commission submitted to the Council a proposal for a Council directive on the approximation of the laws of the Member States restricting the marketing and use of certain dangerous substances and preparations.

This proposed directive contained general restrictive provisions applying to fields which were not covered by other directives, such as those on the composition of petrol (lead content), the sulphur level of fuels, lead and cadmium in ceramics and dangerous substances in paints and varnishes.

In the Commission's view the directive would also have the advantage of allowing more rapid and effective implementation in the European Community of restrictions recommended or adopted by international organizations, such as the decision by the Council on the Organization for Economic Cooperation and Development (OECD) on polychlorinated biphenyls (PCBs).

(a) PCBs

As adopted by the Council on 27 July 1976 the directive applies initially to polychlorinated biphenyls and terphenyls, the use of which might contaminate human health and the environment and to vinylchloride monomer used in aerosols (see (b) below).

The directive specifies that polychlorinated biphenyls, polychlorinated terphenyls and preparations with a PCB or PCT content higher than 0.1% by weight may not be used except for the following categories:

1. Closed-system electrical equipment: transformers, resistors and inductors.

2. Large condensers (1 kg total weight).
3. Small condensers (provided that the PCB has a maximum chlorine content of 43% and does not contain more than 3.5% of penta- and higher chlorinated biphenyls).

    Small condensers which do not fulfil the above requirements may still be marketed for one year from the date of entry into force of this directive. This restriction does not apply to small condensers already in use.
4. Heat-transmitting fluids in closed-circuit heat-transfer installations (except in installations for processing food-stuffs, feeding-stuffs, pharmaceutical and veterinary products. In such installations, the use of heat-transmitting fluids shall still be allowed for 4 years from the date of entry into force of this directive).
5. Hydraulic fluids utilized in:
    (a) underground mining equipment
    (b) machinery servicing cells for the electrolytic production of aluminium in use when this directive is adopted, until 31 December 1979 at the latest.
6. Primary and intermediate products for further processing into other products which are not prohibited under this directive.

A subsequent amendment allowing the use of PCTs in certain thermoplastic tooling compounds has not yet been adopted.

(b) VCM

The directive on marketing and use of dangerous substances adopted on 27 July 1976 also applies to chloro-1-ethylene (vinyl chloride monomer— VCM). It specifies that this substance may not be used as aerosol propellant for any use whatsoever.

At their meeting of 29 June 1978, the EEC Social Affairs Ministers also adopted a Directive on the protection of the health of workers exposed to vinyl chloride monomer.

In view of the threat to workers' health posed by high concentrations of VCM at places where this substance is processed or used, this Directive aims at the adoption of technical preventive and protective measures, based on the latest scientific knowledge, so that the concentrations of VCM in the atmosphere in the works can be reduced to the lowest possible levels. Since there are certain differences in the protective measures adopted by the various Member States, the Directive aims to harmonize and improve existing national laws.

The protection of workers laid down by this Directive comprises:

● technical preventive measures;
● the establishment of limit values for the atmospheric concentration of VCM in the working area;

- the definition of measuring methods and the fixing of provisions for monitoring the atmospheric concentration of VCM in the working area;
- personal protection measures;
- adequate information for workers on the risks to which they are exposed and the precautions to be taken;
- the keeping of a register of workers with particulars of the type and duration of their work and the exposure to which they have been subjected;
- medical surveillance provision (according to the latest medical knowledge).

The provisions of the Directive may be re-examined on the basis of experience grained and in the light of developments in medical techniques and knowledge in this field, the final objectives being to achieve optimum protection of workers.

2. NEW CHEMICAL SUBSTANCES

In addition to the measures discussed above, the Action Programme on the Environment laid down that the Commission would investigate measures still required to harmonize and strengthen control by the public authorities over certain substances or new synthetic products before they are marketed, particularly:

- the improvement and harmonization of quantitative analysis techniques;
- investigations into the long-term toxicity of these substances and the standardization of toxicity tests;
- compulsory submission of samples accompanied by a description of the methods of quantitative analysis.

A preliminary review of the problem conducted on behalf of the Commission indicated that at the present time more than 9,000 synthesized compounds are already used commercially each year in quantities exceeding 500 kg and about 150 compounds are used in quantities exceeding 50,000 tonnes. According to some evaluations, the number of new chemical compounds synthesized every year could rise to about 250,000, a few hundred (500) of which find their way into the various commercial channels. The review pointed out that the use of some of these compounds could have ecological consequences which were sometimes irreversible, if they were put on the market without there being some knowledge beforehand of the potential risk for man and the environment.

The Commission prepared a draft proposal for amending for the sixth time the Council directive of 27 June 1967 relating to the classification, packaging and labelling of dangerous substances.

The proposed amendment provided that the manufacturer or importer was required to carry out a study prior to marketing a new substance to enable its effects on man and the environment to be assessed and to submit to the competent authority at the latest on the date of marketing, a notification with an acknowledgement of receipt including:

- a technical dossier containing all the information necessary to evaluate foreseeable direct or indirect risks which the substance might entail for man and the environment in respect of the various uses envisaged;
- a declaration concerning the unfavourable effects of the substance;
- proposals for the classification and labelling of the substance in accordance with this directive;
- proposals for any measures relating to the conditions of use which were intended to limit the unfavourable effects.

The proposed amendment provided that the person carrying out the notification should send the notification dossier to the competent authority. He may at the same time send a copy of the dossier to the Commission. Details of the technical dossier are given in Table 5.1.

The proposed Directive also provided that Member States should set up or appoint a competent authority (or authorities) who would be responsible for:

- receiving the notification and examining its conformity with the prescriptions of the directive;
- examining the foreseeable risks that these new substances might give rise to;
- examining the classification and labelling proposals;
- examining the proposals of measures relating to the conditions of use.

The proposal laid down that the authorities might, if they saw fit:

- ask for further information and/or verification tests;
- carry out such sampling as is necessary for control purposes;
- take appropriate measures relating to conditions of use while awaiting Community dispositions.

The proposal also laid down a Technical Adaptation Committee procedure where the Commission or the Council might adopt or modify the proposals made for:

- classification;
- labelling;
- dispositions relating to conditions of use.

Table 5.1

INFORMATION REQUIRED FOR THE TECHNICAL DOSSIER ('BASE SET') RE-FERRED TO IN ARTICLE 6(1)

When giving notification the manufacturer or any other person placing a substance on the market shall provide the information set out below.

If it is not technically possible or if it does not appear necessary to give information, the reasons shall be stated.

Tests must be conducted according to methods recognized and recommended by the competent international bodies where such recommendations exist.

The bodies carrying out the tests shall comply with the principles of good current laboratory practice.

When complete studies and the results obtained are submitted, it shall be stated that the tests were conducted using the substance to be marketed. The composition of the sample shall be indicated.

In addition, the description of the methods used or the reference to standardized or internationally recognized methods shall also be mentioned in the technical dossier, together with the name of the body or bodies responsible for carrying out the studies.

1.   IDENTITY OF THE SUBSTANCE

1.1   **Name**

1.1.1   Names in the IUPAC nomenclature

1.1.2   Other names (usual name, trade name, abbreviation)

1.1.3   CAS number (if available)

1.2   **Empirical and structural formula**

1.3   **Composition of the substance**

1.3.1   Degree of purity (%)

1.3.2   Nature of impurities, including isomers and by-products

1.3.3   Percentage of (significant) main impurities

1.3.4   If the substance contains a stabilizing agent or an inhibitor or other additives, specify: nature, order of magnitude: . . .ppm;. . .%

1.3.5   Spectral data (UV, IR, NMR)

1.4.   **Methods of detection and determination**
          A full description of the methods used or the appropriate bibliographical references

2.   INFORMATION ON THE SUBSTANCE

2.1   **Proposed uses**

2.1.1   Types of use
            Describe: the function of the substance ....
                         the desired effects ....

2.1.2   Fields of application with approximate breakdown

            (a) closed system
                — industries ....
                — farmers and skilled trades ....
                — use by the public at large ....

            (b) open system
                — industries ....
                — farmers and skilled trades ....
                — use by the public at large ....

2.2   **Estimated production and/or imports for each of the anticipated uses or fields of application**

2.2.1   Overall production and/or imports in order of tonnes per year 1; 10; 50; 100; 500; 1 000 and 5 000

                — first 12 months .... tonnes/year
                — thereafter .... tonnes/year

2.2.2   Production and/or imports, broken down in accordance with 2.1.1 and 2.1.2, expressed as a percentage

                — first 12 months ....
                — thereafter ....

2.3   **Recommended methods and precautions concerning:**

2.3.1   handling ....

2.3.2   storage ....

2.3.3   transport ....

2.3.4   fire (nature of combustion gases or pyrolysis, where proposed uses justify this)

2.3.5   other dangers, particularly chemical reaction with water

2.4   **Emergency measures in the case of accidental spillage**

2.5   **Emergency measures in the case of injury to persons**
            (e.g. poisoning)

3.   PHYSICO-CHEMICAL PROPERTIES OF THE SUBSTANCE

3.1   **Melting point**

            ....°C

3.2   **Boiling point**

            ....°C.................. Pa

3.3   **Relative density**

            ....$(D_4{}^{20})$

3.4.   **Vapour pressure**

            ....Pa at ....°C
            ....Pa at ....°C

3.5.   **Surface tension**

            .... M/m (..................°C)

3.6   **Water solubility**

            .... mg/litre (..................°C)

3.7   **Fat solubility**
            Solvent — oil (to be specified)
            .... mg/100 g solvent (..................°C)

3.8.   **Partition coefficient**

            n-octanol/water

3.9.   **Flash point**

            ....°C ☐ open cup ☐ closed cup

3.10   **Flammability** (within the meaning of the definition given in Article 2 (2) (c), (d) and (e))

3.11   Explosive properties (within the meaning of the definition given in Article 2(2) (a))

3.12   Auto-flammability

            ....°C

3.13   **Oxidizing properties** (within the meaning of the definition given in Article 2(2) (b))

4.   TOXICOLOGICAL STUDIES

4.1   **Acute toxicity**

4.1.1   Administered orally

          LD_{50}................ mg/kg
          Effects observed, including in the organs ....

4.1.2   Administered by inhalation

          LC_{50}..................(ppm) Duration of exposure .... hours
          Effects observed including in the organs ....

4.1.3   Administered cutaneously (percutaneous absorption)

          LD_{50}..................mg/kg

          Effects observed, including in the organs ....

4.1.4   Substances other than gases shall be administered via two routes at least, one of which should be the oral route. The other route will depend on the intended use and on the physical properties of the substance.

          Gases and volatile liquids should be administered by inhalation (a minimum period of administration of four hours).

          In all cases, observation of the animals should be carried out for at least 14 days.

          Unless there are contra-indications, the rat is the preferred species for oral and inhalation experiments.

          The experiments in 4.1.1, 4.1.2 and 4.1.3 shall be carried out on both male and female subjects.

4.1.5   Skin irritation

          The substance should be applied to the shaved skin of an animal, preferably an albino rabbit.

          Duration of exposure................... hours

4.1.6   Eye irritation

          The rabbit is the preferred animal.

          Duration of exposure .... hours

4.1.7   Skin sensitization

          To be determined by a recognized method using a guinea-pig.

4.2   **Sub-acute toxicity**

4.2.1   Sub-acute toxicity (28 days)

          Effects observed on the animal and organs according to the concentrations used, including clinical and laboratory investigations ....

          Dose for which no toxic effect is observed ....

4.2.2   A period of daily administration (five to seven days per week) for at least four weeks should be chosen. The route of administration should be the most appropriate having regard to the intended use, the acute toxicity and the physical and chemical properties of the substance.

Unless there are contra-indications, the rat is the preferred species for oral and inhalation experiments.

4.3   Other effects

4.3.1   Mutagenicity (including carcinogenic pre-screening test)

4.3.2   The substance should be examined during a series of two tests, one of which should be bacteriological, with and without metabolic activation, and one non-bacteriological.

5.   ECOTOXICOLOGICAL STUDIES

5.1   Effects on organisms

5.1.1   Acute toxicity for fish

LC_{50} .... (ppm) Duration of exposure determined in accordance with Annex V (C)

Species selected (one or more) ....

5.1.2   Acute toxicity for daphnia

LC_{50}................. (ppm) Duration of exposure determined in accordance with Annex V (C)

5.2   Degradation

— biotic

— abiotic

The BOD and the BOD/COD ratio should be determined as a minimum

6.   POSSIBILITY OF RENDERING THE SUBSTANCE HARMLESS

6.1   For industry/skilled trades

6.1.1   Possibility of recovery ....

6.1.2   Possibility of neutralization ....

6.1.3   Possibility of destruction:

— controlled discharge ....

— incineration ....

— water purification station ....

— others ....

6.2   **For the public at large**

6.2.1   Possibility of recovery ....

6.2.2   Possibility of neutralization ....

6.2.3   Possibility of destruction:

          — controlled discharge ....

          — incineration ....

          — water purification station ....

          — others ....

The proposal was adopted on 18 September 1979. As adopted, the directive obliges Member States to take all measures necessary to ensure that substances are not placed on the market before being notified to a competent authority. The notification must be sent 45 days before marketing in the form of a technical dossier conforming to the requirements of the directive. Further information may be requested by the competent authority in accordance with requirements set out in the directive. Exceptions are provided for substances manufactured in quantities of less than 1 tonne per year, and those used for research and development. A copy of the dossier is to be sent to the Commission who will pass it on to other Member States. There are provisions concerning confidentiality and the setting up of an inventory by the Commission.

On 31 March 1982 the Commission of the European Communities published in the Official Journal: the European Core Inventory (ECOIN) of approximately 34,000 commercial chemical substances known by the Commission to be on the EC market before 18 September 1981.

This inventory is required under the 6th amendment of the 1967 Directive on the Classification, Packaging and Labelling of Dangerous Substances (Council Directive 79/831/EEC).

All commercial chemicals not included in the final European Inventory of Existing Chemical Substances (EINECS) must be notified to a Member State before they are marketed in the European Community.

Publication of the ECOIN is a major step towards completing the European Inventory of Existing Chemical Substances which ultimately will list those chemical substances not subject to new chemical notification.

The Commission has had a number of discussions with the United States regarding the implementation of the Toxic Substances Control Act (TSCA), following a Decision by the Council to this effect of 30 May 1978.

3. PLANT PROTECTION PRODUCTS

On 4 August 1976 the Commission submitted to the Council a proposal for a Council directive concerning the placing of EEC-accepted plant protection products on the market. The use of plant protection products is essential in modern agriculture for the protection of crops and crop products from the effects of harmful organisms and weeds. They contribute thereby to the improvement of the productivity of agriculture and to assuring the availability of supplies.

Many plant protection products contain active substances whose use can also present risks to man, animals, plants and the environment and most Member States have rules governing both the marketing and use of such products. These rules differ and particularly those relating to marketing may be an obstacle to the free movement of goods within the Community and may also constitute a disincentive to innovation.

The proposal for a directive deals with the marketing and otherwise placing on the market of plant protection products.

It envisages the creation of an optional 'EEC-acceptance' to operate in parallel with existing national arrangements for approving plant protection products. An applicant wishing to market a plant protection product within the Community will have the choice either to apply for separate registration under national legislation as at present or to apply for EEC-acceptance, on its own or in addition to national registrations, to one of the Member States in accordance with the provisions of this directive. EEC-acceptance, if granted, would be recognized, subject to certain safeguards, by all Member States normally within a 1–2 year period.

EEC-acceptance will permit the free circulation of the product throughout the Community except in so far as Member States may be authorized, particularly because of local conditions, to prohibit its circulation in their territory or to restrict or vary its field of use.

Such a directive was called for specifically in the Council Resolution of 22 July 1974 on the veterinary, plant health and animal feeding-stuffs sectors. It also represents a contribution to the Communities' environmental policy as envisaged in the Programme of Action of the European Communities on the Environment.

This proposal complements the proposal for a Council directive on the approximation of the laws, regulations and administrative provisions of Member States relating to the classification, packaging and labelling of pesticides adopted by the Council on 26 June 1978. Together with the latter, it would considerably improve the protection given to the users of plant protection products and to consumers of plants and plant products.

It will be noted that this proposal does not provide for total harmonization. The market for plant protection products in the Community is specialized and characterized by the large number of products available. Many thousands of different commercial preparations containing several hundred active substances are in use in the Community—in France alone over 6,000 products are officially approved for use in agriculture, although admittedly by no means all are of economic importance. Furthermore, a considerable proportion of nationally registered plant protection products is intended only for local or regional marketing to meet local or regional agricultural and ecological conditions and needs, which can vary significantly over a geographical area as large as the Community. Under these circumstances, it has been judged desirable at this stage to permit Member States to continue to approve for marketing in their own territory plant protection products in accordance with national provisions. In this way the proposal offers the necessary flexibility for manufacturers and distributors of products with only limited regional application.

The proposal has not been adopted by the Council.

On 4 August 1976 the Commission sent to the Council a complementary proposal for a Council directive prohibiting the placing on the market and use of plant protection products containing certain active substances. The proposal for a directive concerns prohibition of the marketing and use of certain plant protection products containing active substances whose use is hazardous to human or animal health or is unduly harmful to the environment.

In a first stage, it covers plant protection products containing certain mercury or organo-chlorine active substances. These substances have been the subject of extensive study in recent years, their use is generally recognized to be undesirable in agriculture and has already been prohibited, or otherwise discontinued, in some Member States. The directive was adopted on 21 December 1978.

4. DANGEROUS INDUSTRIAL ACTIVITIES

The control of major hazards related to dangerous industrial activities still is an extremely complex problem, despite the considerable efforts in the various countries of the Community. Accidents like those of Flixborough in 1973 and of Seveso in 1976 are only two of the examples which tragically prove that safety problems of chemical plants need new solutions, more in line with today's technical and scientific knowledge.

The Commission prepared a draft Directive on the control of certain industrial activities with regard to accidental hazards. Discussions on this draft started in April 1977 within a group of national experts which met five times. It was also discussed by the European industry within the working of CEFIC (Conseil Européen des Fédérations de l'Industrie Chimique). Before transmission of the draft to the Council, the opinion of the tripartite Advisory Committee on Safety, Hygiene and Health Protection at work was sought with regard to the problems of protection of workers.

The proposal for a directive covered industrial activities which involved or might involve the categories of dangerous substances defined in the 1967 dangerous substances directive, i.e. substances which are explosive, oxidizing, easily flammable, toxic, harmful, corrosive, irritant and hazardous to the environment. It applied to new industrial activities, any alterations with implications for the safety of an industrial activity, and existing industrial activities for which certain periods of grace would be allowed. The first part had the role of a framework directive to cover any industrial activity, including storage conditions which involved or might involve dangerous substances. In general it provided for taking the measures needed to prevent accidents and to limit the consequences of accidents which did occur. The second part applied more particularly to industrial activities which involved or might involve particularly dangerous substances clearly defined in a list and by means of criteria in the directive which were present or potentially present in the industrial activity in excess of a particular quantity. To monitor the safety of industrial activity, manufacturers would be required to notify competent authorities a detailed safety report concerning the substances, facilities and points where major accidents could occur. The proposal also provided for competent authorities to be notified of major accidents and for the Commission to establish a data bank relating to accident hazards and major accidents which had actually occurred.

The directive was agreed on 3 December 1981. Agreement had been held up for a long period over the procedure to be followed where accidents could have effects across frontiers. As adopted on 24 June 1982 the directive applies to specified industrial activities (including storage) both new and existing which could present major accident hazards. It does not apply to nuclear installations, military installations, manufacture and storage of explosives, mining operations and installations for the disposal of toxic and dangerous waste covered by other Community legislation. It requires Member States to ensure that a manufacturer involved in any of the specified activities takes all necessary measures to prevent major accidents and to limit their consequences. In the case of an activity which is concerned with certain specified dangerous substances over a certain quantity (including storage) a notification scheme is to be set up to ensure that all appropriate information is provided to a competent authority. If an accident occurs, certain other notification requirements must be observed, and information passed to the Commission. Persons liable to be affected by a major accident are required to be informed of the safety measures; where the effects could concern persons in other Member States, the same information is to be provided to these states as a basis for consultation within their own bilateral relations. A statement in the Official Journal specifies the obligation of Member States to consult in these terms. The directive enters into force on 8 January 1984 but need not apply to existing plants until 8 January 1985.

5. ECDIN AND SCIENTIFIC ADVISORY COMMITTEE

(a) ECDIN

The EEC has set up a data bank named ECDIN (Environmental Chemicals Documentation and Information Network), a pilot project which since

1972 has been testing on an experimental basis an embryo data bank and public information service on chemical products potentially dangerous to man or the environment. The ECDIN project has reached a stage of development where it is possible to consider moving to an operational phase.

ECDIN is designed as an information service providing any public body or private undertaking with fast, detailed, up-to-date and accurate information on any chemical produced in large quantities and liable to have harmful effects on human beings, animals or the environment.

It is estimated that not only should 20,000 to 30,000 chemicals be catalogued (nomenclatures exist) but also that a description should be given of the toxicity, the cycles and quantities in which they are produced and used, their dispersion capacity, their passage in the biotic and abiotic environment in a series of different forms, and the known methods of detection, prevention and intervention.

At present, in the pilot phase, ECDIN contains information on some 5,000 compounds broken down into ten sectors: identification (chemical standard, commercial standards, etc.); chemical structure; physical-chemical properties; analysis/data; production and sale; carriage, packaging, danger; utilization and elimination; distribution and conversion in the environment; toxicity, ecotoxicity; legislation.

The Joint Research Centre manages the data bank as a system to fulfil the needs and requests of the Commission in application of the various directives and draft directives concerning chemicals released into the environment. The pre-operational phase still needs research and development essentially on software and data validation. Part of this is being carried out under the indirect action programme.

(b) Scientific Advisory Committee on Ecotoxicity

In approving the Second Action Programme on the Environment, the Council recognized that there should be a systematic review on the basis of toxicity, bio-accumulation and persistence criteria, of the specifications and manner of use of certain chemical compounds meeting those criteria, having regard to:

- the better scientific understanding of ecotoxicity acquired since they were first put on the market;
- more accurate assessment of the exposure levels of targets;
- new uses to which the chemical compounds in question are put;
- their compatibility with recycling techniques.

The Commission will conduct this review with the assistance of a committee of national experts. (Commission Decision of 28 June 1978.)

6. FUTURE WORK

The Commission's proposals for the 3rd Action Programme provide for:

(a) the setting up by the Commission of a coordinating committee, consisting of representatives of the Member States, to supervise the monitoring of chemical substances and to facilitate close cooperation between competent national authorities particularly in applying the Sixth Amendment to the 1967 Directive on dangerous substances;

(b) continuation of the Commission's work on updating the directives already adopted on dangerous substances and preparations.

(c) harmonization of the assessment of the impact of chemicals on man and the environment to avoid differing assessments in the various Member States, which would be detrimental to the proper functioning of the common market: at regular intervals lists of existing dangerous substances which require priority assessment will be drawn up by the Commission, who will be assisted in this task by the Scientific Advisory Committee to examine the toxicity and eco-toxicity of chemical compounds, which was set up in 1979.

(d) intensification of measures which can lead to international agreements on toxic substances either bilaterally with certain non-member countries or internationally through, say, the OECD.

6

Noise

The first Programme of Action of the European Communities on the Environment, which the Council adopted on 22 November 1973, did not include a chapter specifically headed 'Noise'. The Council did not adopt, nor did the Commission propose, a coherent programme in this field. The actions envisaged related essentially to harmonization activities in implementation of the general programme for the elimination of technical barriers to trade. In order to remove the economic distortions which are liable to be created by differences in specifications for certain products and noisy equipment, the Commission was asked to submit to the Council a number of proposals for directives. At the same time the Council recognized that the fact of harmonization, though basically motivated by trade and economic considerations, could be used to serve an important environmental goal. It was the Council's clear intention, when it adopted the first Environment Programme, that the general programme for the elimination of technical barriers for trade should also serve the Community as a tool to achieve a general improvement in the quality of the environment.

1. MOTOR VEHICLE NOISE

(a) 1970 directive

The first priority was motor vehicle noise. It has long been accepted that noise is one of the major nuisances which go hand in hand with urban development and that the noise of motor vehicle traffic is generally considered to be the chief offender. There have been some very thorough surveys carried out in cities such as Chicago, London, Paris, Nice and New York to assess the amount of discomfort that people experience when confronted with different noises. The results published in the report, for example, by Professor Wilson when he was Chairman of the Research Cooperation Committee of the OECD, show that traffic noise comes top of the list with 36% of the people concerned, followed by aircraft noise with 9% and noise from railway trains with 5%.

In fact, most countries in Europe introduced measures to restrict motor vehicle noise soon after the war. Of course, these measures were not standardized and specialists at ISO (the International Organization for Standardization) felt that a standard ought to be worked out to lay down measuring methods and vehicle operating conditions which would enable precise and reproducible results to be obtained.

Work began in July 1958 and a draft standard was drawn up in 1960. After amendments had been made, it was put to the vote of the Member Bodies in May 1962. It was approved by 27 countries with only one country against, and was formally published in February 1964 as ISO Recommendation R 362/Measurement on Vehicle Noise.

At that time several European countries adopted it as their official method of measurement and fixed maximum sound levels for the various vehicles categories. When the Commission of the European Communities came to examine the problem of vehicle noise, it also drew on ISO Recommendation R 362 to draw up the draft directive adopted by the Council of the European Communities on 6 February 1970.

This Directive (70/157/EEC) on the approximation of the laws of the Member States relating to the permissible sound level and the exhaust system of motor vehicles is now in force in all motor vehicles intended for use on the road, having at least four wheels, with the exception of agricultural tractors and machinery and civil engineering equipment.

(b) Further Commission proposals on motor vehicle noise

It is fair to say that the application of the provisions of the Council directive of 6 February 1970 has already led to a significant reduction of noise nuisance in urban centres of the Community. In France for example, since the early 1960s noise levels have dropped by 7–10 dB(A) in the case of commercial vehicles and 6–8 dB(A) in the case of passenger vehicles. However, the continued increase in the number of vehicles on the road has rendered these provisions partially ineffective. This situation, together with increased demands to protect the urban population and the environment against noise nuisance, has led the Commission to conclude that the limits permitted in the first directive must be made more stringent.

On 20 June 1973 and 5 September 1973 respectively, the Governments of France and the United Kingdom informed the Commission of their interest in seeing a substantial reduction of the present limits and invited the Commission to examine the corresponding possibilities.

After consultations with national experts, the Commission drew up a proposal for a Council directive modifying the Council directive of 6 February 1970. In view of the complexity of the subject this proposal contained both short-term and long-term elements.

The *short-term programme* aimed at an initial reduction in the present limits without changing fundamentally the test method and in accordance with the same vehicle classification scheme as laid down in the directive already adopted. The proposed reductions, however, took account of the need to allow certain modifications to be made to the test method, the effect of which would be to make the specifications for some types of car even more stringent. For this reason the reduction proposed for other types of vehicle was greater, as much as four decibels for buses, i.e. a reduction of about 50%. Indeed, the Commission considered that it was possible to further reduce the limits for these vehicles, in view of their major contri- bution to urban traffic noise and also with a view to reducing the gap between the limits permitted for the various categories of vehicle.

In the Commission's view, this reduction for buses was no more than a first step and is to be followed by further significant reductions aimed at keeping the sound emissions from these vehicles to a minimum; this will be technologically feasible as soon as certain design features are widely adopted (see Table 6.1).

The aim of *the longer-term programme* is to find a new method of noise measurement which reflects the actual conditions in which vehicles are used in urban traffic and will, on the basis of the results of current studies and research and in accordance with the procedure laid down for adaptation to technical progress, lead to further amendments to the directive of 6 February 1970.

The Council adopted the proposed amendment to the 1970 directive on 8 March 1977. At the same time it made a declaration as follows:

> efforts should be made to achieve a noise level of around 80 dB(A) for all categories of vehicle by 1985. The levels decided on will have to take into account what is technically and economically feasible at the time. Moreover, they will have to be established sufficiently early to give manufacturers an adequate transition period in which to improve their products.

Details of the levels set in the 1970 directive and the 1977 amendment are at Table 6.1.

The directive is an optional one i.e. it does not impose limits on member states but ensures that no member state may refuse to grant EEC type approval or national type approval on grounds relating to the permissive sound level if the sound level meets the limits of the directive.

An adaption to the directive was adopted on 13 April 1981 introducing a new test procedure in October 1984. Though the numerical limits remain unchanged the noise limits of some vehicles will effectively be reduced by up to 3 dB(A).

(c) Tractor noise

An 'optional' directive setting noise limits for agricultural tractors was adopted by the Council on 28 March 1974. The limits set were 89 dB(A)

Table 6.1: Council Directive of 6 February 1970 and
amendment of 8 March 1977
Permissible sound level for motor vehicles

Vehicle category	Value expressed in dB(A) 1970	Value expressed in dB(A) 1977
1.1.1. Vehicles intended for the carriage of passengers and comprising not more than nine seats including the driver's seat	82	80
1.1.2. Vehicles intended for the carriage of passengers, comprising more than nine seats including the driver's seat, and having a permissible maximum weight not exceeding 3.5 tonnes	84	81
1.1.3. Vehicles intended for the carriage of goods and having a permissible maximum weight not exceeding 3.5 tonnes	84	81
1.1.4. Vehicles intended for the carriage of passengers, comprising more than nine seats including the driver's seat, and having a permissible maximum weight exceeding 3.5 tonnes	89	85
1.1.5. Vehicles intended for the carriage of goods, and having a permissible maximum weight exceeding 3.5 tonnes	89	86
1.1.6. Vehicles intended for the carriage of passengers, comprising more than nine seats including the driver's seat, and having an engine power equal to or exceeding 200 HP DIN	91	87
1.1.7. Vehicles intended for the carriage of goods, having an engine power equal to or exceeding 200 HP DIN and a permissible maximum weight exceeding 12 tonnes	91	88

for tractors weighting more than 1.5 tonnes (unladen) and 85 dB(A) for tractors weighing less than 1.5 tonnes.

2. MOTORCYCLE NOISE

On 12 December 1975 the Commission submitted to the Council a proposal for a Council directive on the approximation of the laws of the Member States relating to the permissible sound level and to the exhaust system of motorcycles.

In making its proposal, the Commission recognized that motorcycles represented a by no means negligible proportion of the motor vehicles on the roads in urban centres. Moreover, this type of vehicle was tending to increase in number. In view of their technical characteristics, motorcycles were frequently used at full throttle, i.e. at very high engine speeds. It has been proved that with most motorcycles a 20% increase in speed causes at least a doubling of the sound intensity emitted.

Table 6.2: Council Directive of 12 December 1975
Proposed Permissible Sound Levels for
Motorcycles

Category of cubic capacity (cm³)	Permissible sound level db(A)
≤ 80	78
≤125	80
≤350	83
>500	85
500	86

The scope of the directive is limited to two or three wheeled motorcycles whose maximum designed speed is greater than 45 km/h. Provisions relating to the permissible sound level and to the exhaust system of such motorcycles are integrated into a EEC type-approval procedure, similar to that already established in the case of motor vehicles. Technical annexes contain the requisite definitions, the procedure for application for EEC type-approval, the limits to be observed as regards the sound level during the testing of motorcycles in motion and the requirements concerning measuring instruments, conditions and methods.

This directive is also an 'optional' one.

The proposed directive was adopted by the Council on 16 October 1978. Details of permissible sound levels are given in Table 6.2. The directive provides for the Council to decide on a further reduction in the maximum noise of motorcycles by 31 December 1984.

3. CONSTRUCTIONAL PLANT AND EQUIPMENT

Acting on a proposal from the Commission, the Council included the sector 'constructional plant and equipment' in the supplement of 21 May 1973 to the general programme of 28 May 1969 aimed at eliminating technical obstacles to trade in industrial products. The action programme on the environment also called for the Commission to submit proposals to the Council in this field. On 20 December 1974, the Commission sent to the Council proposals for Council directives dealing both with the noise level of constructional plant and equipment and also with the measurement method.

For the purposes of the directives, 'constructional plant and equipment' means all machinery, appliances, equipment and installations which are used, according to their type to perform work on building sites, but are not primarily intended for the transport of goods or persons. Like the motorcycle directive, the proposal follows the 'optional' solution.

The European Parliament, though expressing itself in general in favour of the proposals, has once again commented that the optional type of harmonization laid down hardly seems compatible with the dangers to health and the serious effects on the environment caused by this kind of equipment.

The proposal to establish procedures has not yet been agreed by Council but that on measurement was adopted on 19 December 1978 and amended on 7 December 1981. In the meantime the Commission has submitted specific proposals under the general directive for permissible sound levels for the following types of equipment on the dates shown:

- concrete breakers and jackhammers—31 December 1974
- tower cranes, current generators for welding and for power supply—30 December 1975
- air compressors—5 April 1978
- dozers, loaders and excavators—28 November 1980
- excavator loaders—9 October 1981

A proposal on mobile cranes is in preparation.

In the case of tower cranes and current generators, the 'information agreements' have served a useful purpose. The German Government notified the Commission of a draft regulation providing for progressive limitation of the sound level of tower cranes and the French Government informed the Commission of two draft orders relating to the limitation of airborne noise emitted by current generators for power supply and welding.

In making these proposals, the Commission recognized that the ISO has for some years been studying draft standards for noise emitted by constructional plant and equipment. The Commission endeavours to achieve the closest possible collaboration with the ISO, both at the professional and personal level. The Commission recognizes that the trade in constructional plant and equipment manufactured in the Community countries far transcends the frontiers of the Community. It should certainly be possible to consider applying the principle of strict reference to ISO standards in the event of such standards being published between now and the time these directives are adopted by the Council. In this case, the framework directive provides for the adaptation to technical progress of the general test method and also the general survey method, as well as the technical annexes of the special directives.

4 AIRCRAFT NOISE

(a) Work of ICAO

Another area where the Community is seeking to take action on the source of noise itself is aircraft. The Environment Programme of 22 November

1973 did not mention aircraft as such. However, the Council, in reply to Written Question No. 654/73 put by Members of the European Parliament on the subject of aircraft noise, stated that 'the Environment Programme of the European Communities provides for mounting a campaign against environmental and noise pollution caused by aircraft'. In that reply, the Council also envisaged standards for aircraft, making use of work done by international organizations.

The Fifth Air Navigation Conference of the International Civil Aviation Organisation (ICAO) in 1967 made certain recommendations based on the principal conclusions of the International Conference on the reduction of noise and disturbance caused by civil aircraft (London 1966) with the object of reaching international solutions to the problem through the ICAO. This led the ICAO Council on 2 April 1971 to adopt the first set of standards and recommended practices on aircraft noise, known as 'Annex 16 to the International Convention on Civil Aviation'.

Among other things Annex 16 contains international standards and recommended practices relating to the noise certification of various categories of aircraft.

The first amendment to Annex 16, adopted by the ICAO Council on 6 April 1973, became operative on 16 August 1973. It covered subsonic jet aircraft with a maximum take-off weight greater than 5,700 kg, powered by engines with a by-pass ratio greater than two and which received their first individual certificate of airworthiness after 1 March 1972, or aircraft powered by other categories of engine and which were granted their type certificate of airworthiness after 1 January 1969. The same standards applied to subsonic jet aircraft with a maximum take-off weight exceeding 28,500 kg and powered by engines with a bypass ratio less than two if the type certificate of air-worthiness was issued before 1 January 1969 and their first individual certificate of airworthiness was not issued before 1 January 1976.

The second amendment adopted by the ICAO Council on 3 April 1964 came into force on 27 February 1975. It extended the scope of the Annex to include all recent jet aircraft, irrespective of weight, and introduced recommended practices for the noise certification of light propeller aircraft.

(b) Commission proposal on noise from subsonic aircraft

On 2 April 1976, the Commission made a proposal for a Council directive on the limitation of noise emission from subsonic aircraft. In the light of the developments at ICAO, and of progress at international level the Commission believed that the most effective way of reducing aircraft noise was by the uniform application in all the Member States of the European Community of ICAO standards, in particular those set out in the latest version of Annex 16 to the Chicago Convention on International Civil Aviation. Article 37 of this Convention, of which Annex 16 is a part, requires each contracting State to undertake to achieve the greatest possible uniformity in regulations and standards. The latter do not become mandatory in a State until embodied in its national laws.

In making its proposal the Commission noted that the relevant laws of most Member States of the European Community were based on the principles of Annex 16. Nevertheless, there were major discrepancies between them. Italy, for example, had at present only a code of practice which embodied in the 'Registro Aeronautica Italiana' and provided for a system of certification based on the standards of Annex 16, and Luxembourg still had no laws at all in this field.

Of the other Member States, Germany, France, Ireland, the Netherlands and the United Kingdom based their laws on the first amendment to Annex 16.

Belgium and Denmark based their laws on the latest version of Annex 16.

In the Commission's view such discrepancies not only ran the risk of limiting the effectiveness of the measures to combat aircraft noise, but also of creating distortions in competition between purchasers (airline companies) which would have a direct effect on the functioning of the common market.

The draft directive embodied the requirements of the latest version of Annex 16 to the Convention on International Civil Aviation, adopted by the ICAO Council on 3rd April 1974 and operative from 27 February 1975.

The directive would establish an EEC noise limitation certificate, i.e. the document by which the Member State which has registered the aircraft recognizes that such aircraft meet the requirements of the directive.

The proposal provided for checks on compliance and also for the exchange of information between Member States to make such checks easier. Clearly, actions against aircraft noise have to be on an international scale. The directive therefore not only concerns aircraft on the civil aviation registers of Member States. It also related to all civil aircraft of non-member countries landing or taking off in a Member State. In accordance with Article 37 of the Convention on International Civil Aviation, the proposal would impose the ICAO standards and recommended practices on aircraft of non-member countries.

The proposed directive would require mutual recognition of an EEC noise limitation certificate issued by a Member State. It left Member States free to impose restrictions on aircraft outside the scope of this directive. It therefore provided an incentive to modernize fleets and consequently reduce noise emission.

The directive on aircraft noise was adopted by the Council on 20 December 1979.

As adopted the directive makes mandatory throughout the community the noise standards which had previously been agreed within the International Civil Aviation Organisation. It also makes mandatory the recommendations of the European Civil Aviation Conference (ECAC) concerning the use of non-certificated aircraft. Each Member State is responsible for ensuring that aircraft registered in its territory in the categories covered may not be used in the territory of other Member States unless granted a noise certification complying with the standard laid down.

(c) Proposed amendment to the directive on noise from subsonic aircraft

On 28 September 1981 the Commission submitted a proposal for amending the directive on limitation of noise from subsonic aircraft of 20 December 1979. The changes proposed arise in the first place from amendments to Annex 16 of the Convention on International Civil Aviation adopted by the Council of ICAO on 11 May 1981. They are technical in nature and introduce improvements in the noise certification requirements for conventional propeller-driven aeroplanes and subsonic jet aeroplanes and introduce units of measurement in System International (SI). It is also proposed to require Member States to prevent non-noise-certificated subsonic jet aircraft not registered in a Member State operating in their territories from 1 January 1988.

(d) Noise from helicopters

On 13 October 1981 the Commission submitted a proposal for a directive on the limitation of noise emissions from helicopters. An amendment to Annex 16 to the Convention on International Civil Aviation approved by the Council of ICAO on 11 May 1981 introduced standards for the noise certification of helicopters. These became applicable internationally on 26 November 1981. The proposal would make these standards mandatory in Member States in the same way as required by the directive on noise emission from subsonic aircraft.

(e) Relations with international bodies

On 12 June 1978, EEC Transport Ministers decided that the Community priority in the field of air transport should be the setting of noise standards for aircraft. It also recognized the importance of establishing links with important international organizations such as the European Civil Aviation Conference (ECAC) and the International Civil Aviation Organization (ICAO).

5. ACTION IN OTHER AREAS

(a) Lawnmowers

On 18 December 1978 the Commission submitted a proposal for a directive

concerning noise emissions from lawnmowers. This proposal had its origin
in legislation notified under the 'information agreement'. The proposal
provided for noise limits for motorized lawnmowers in an initial stage with
proposals for further reductions to follow. Total harmonization was proposed. The proposal (along with others in the elimination of technical
barriers to trade programme) has not yet been agreed.

(b) Domestic appliances

On 19 January 1982 the Commission submitted to the Council a proposal
for a directive concerning airborne noise emitted by household equipment.
It had its origin in a notification from the French Government that they
intended to adopt administrative provisions to limit noise from electric
household appliances. The German and Dutch Governments are preparing
similar measures. The proposal has two objectives—one as part of the
Community Environment Programme, to combat noise; and, two, to remove technical barriers to trade. The Commission felt that as a first step
the best way to achieve this two-fold aim in the highly diversified industry
would be to call on producers to tell purchasers the level of noise emitted
by their equipment rather than to impose specific noise limits. The proposal,
therefore, lays down detailed rules on the publication of noise levels along
with the general principles on which to base the noise measurements and
the broad lines of the method of checking the accuracy of the figures given.
It would be a framework type of directive requiring implementing directives
for each family of domestic appliance e.g. washing machines, dishwashers.

(c) Noise from trains

The Commission have in preparation a proposal concerning noise from
trains.

(d) Acoustic standards in housing

The Commission has also been doing work on a proposal for a directive
for acoustic standards in housing but further research is needed.

6. CRITERIA

In the language of the Communities Environment Programme the term
'criterion' signifies the relationship between the exposure of a target to
pollution or nuisance and the risk and/or the magnitude of the adverse or
undesirable effect resulting from exposure in given circumstances.

Meetings of national experts have been held to discuss and to critically
analyse the available bibliography on the adverse or undesirable effects of
the exposure of man to noise. The results of this work are given in *Damage*

and Annoyance caused by Noise (Document No EUR5398e, rapporteurs: H. Bastenier, W. Klosterköter and J. B. Large).

The Commission has also taken into account the work performed at national and international level. In particular it has considered the report published by the WHO (Geneva) in 1973, entitled *Health Hazards of the Human Environment*, as it relates to noise.

On 3 December 1976 the Commission sent to the Council a Communication concerning the determination of criteria for noise in respect of five categories:

(a) sleep interference
(b) speech interference
(c) annoyance
(d) performance of tasks
(e) hearing damage

The essential elements of this Communication are given in Table 6.3.

7. FUTURE WORK

In adopting the Second Action Programme on the Environment, the Council stated that a consequence of the overall increase in noise was the need in future for solutions to this general problem to be sought through the implementation of an overall anti-noise programme, initiated in general outline at Community level, and worked out in detail and implemented at Community, national, regional or local level, according to the type of measure intended.

The Commission would propose a programme as soon as possible setting out the general framework for a body of measures which should be taken at these different levels to combat noise. The measures concerned must be specified and varied according to the types of activity that it would be desirable either to protect from noise (activities such as education, medical care, relaxation, rest, leisure, etc.) or to regulate in order to reduce the noise they cause (transport, industry, agriculture, noisy leisure activities, etc.). The measures should not only cover the sources of noise emission but should also take into account the conditions governing noise propagation and reception (e.g. traffic noise can be limited not only by reducing engine noise levels, but also by the use of better road surfacing materials and by siting roads with greater care).

This general proposal could give rise, depending on circumstances, to Community action or to national or regional programmes, taking into account any special economic and social features.

With regard to Community action, the Commission would make appropriate proposals concerning:

Table 6.3: Commission Communication of 3 December 1976
Determination of noise criteria

(a)	*Sleep interference*
—	35dB(A) is the non-fluctuating (less than 5dB) continuous indoor level up to which reports concerning sleep disturbance or awakenings are constant from about 10% of the subjects tested irrespective of the cause.
—	The continuous equivalent indoor level above which the pattern of sleep (e.g. EEG) is changed in more than 10% of the subjects and above which the percentage of reports is significantly increased is between 40dB(A) (about 20% reports) and 50dB(A) (about 50% reports).
—	Changed activation of the central nervous system, which may lead to awakening, is observed if an increase of 10dB or more occurring in 0.5 sec. or less is superimposed on a continuous background level.
—	Reduced sleeping ability of the particularly sensitive population (e.g. old, sick, convalescent) has been demonstrated at values approximately 10dB below those mentioned above.
(b)	*Speech interference*
—	A continuous equivalent level of 65dB(A) makes normal conversation just possible at 1 metre.
—	A continuous equivalent level of 45dB(A) or less introduces no problems in normal conversation at a distance of 1 metre. At greater distances lower levels correspond to these effects on speech intelligibility.
—	In special situations where the contents of the message have to be completely understood, for example teaching in classrooms, medical consultations, the levels of background noise corresponding to the levels in paragraphs 2.1 and 2.2 should preferably be about 10dB lower.
—	For television viewing, listening to the radio, or telephone conversations, in cases where the background noise levels show large variations with time, equivalent noise levels corresponding to the levels in paragraphs 2.1 and 2.2 are about 5dB lower.
(c)	*Annoyance*
—	Under average town living conditions, outside noise which emanates from transportation and industrial sources of 45dB(A) equivalent daytime noise levels will generally cause about 15% of the population to be highly annoyed. 65dB(A) will generally cause about 40% of the population to be highly annoyed.
—	In noisier living conditions, city centres and near industrial sites, somewhat higher noise levels will correspond to the above described effects. On the other hand in quieter situations such as rural areas the described effects will occur at correspondingly lower levels. Tones and impulsive noises present in the environment increase the level of annoyance at each value of the equivalent noise level.
—	During periods where the sensitivity to noise is greater, such as periods of rest or relaxation, the corresponding noise levels are lower.
(d)	*Performance of tasks*
	The findings of laboratory work show in general:
—	A steady noise, without special significance, would not appear to interfere with most human activities that do not require acoustic information in order to be carried out. This is so even where the steady level is relatively high, possibly as high as 90dB(A) at some times.

Table 6.3 continued

— Intermittent or impulsive noise has a more marked disturbing effect than steady noise.
— High frequency noise components (above about 200 Hz) usually cause worse interference with performance than do low frequency components.
— Noise does not have a notable effect on overall performance—but high levels of noise can cause variations in the performance of sequential tasks. There can here be a complete breakdown of performance or a total absence of reaction to stimuli, sometimes followed by a compensating improvement.
— Noise affects the quality of work more than the quantity.
— Complicated tasks, demanding considerable concentrations, are more easily influenced by noise than simple tasks.
(e)   *Hearing damage*
High level noise can cause permanent impairment to hearing, different to age and illness, and which can lead to a handicap. Such a handicap can be avoided for the great majority of the population if the noise level to which they are exposed over their whole lifetime is less than a 24 hour daily value of equivalent continuous sound level, Leq., of 80dB(A); higher levels of continuous noise may only be endured, with no damage occurring, for shorter periods of time. The damage risk is greater if impulsive noise is added to continuous noise—the human ear cannot tolerate without damage, noise having an instantaneous value greater than 150dB(A).
In the determination of these criteria, it was considered that specific noise sources such as aircraft, traffic, musical noise etc. were likely to produce different levels of acceptability. The exposure/effect curves for each type of noise are not identical.
It was also considered that because of interindividual differences between members of the general population, the effects of noise can vary considerably.
In the determination of these criteria the population as a whole was considered. Additional considerations should be given to hypersensitive sections of the population, for example, old people, the sick and the very young.
This proposal concerning the determination of criteria for noise does not cover vibrations, and subsonic and supersonic waves, which will be the subjects of further studies.

- the guidelines which the competent authorities may take into consideration when determining the levels (quality objectives) appropriate to zones, where a particular activity predominates: rest zones, residential areas, leisure areas, industrial estates, roads, railways, airports, international waterways, etc.;
- noise measurement methods;
- specifications for noisy products, possible measures dealing with the monitoring of the utilization of these products, rules for labelling and the affixing of stickers; with the assistance of national experts the Commission will draw up a list of priorities with a view to tabling proposals on these matters. Such a list should be based on an assessment of the contribution of such apparatus to the overall impact of noise on the environment;

- noise-insulation standards;
- permissible noise levels at the place of work, in conjunction with the European Foundation for the Improvement of Living and Working Conditions and with the Social Action Programme of the European Communities.

The Commission would also carry out research into the little-known effects of noise on man (especially short-lasting and low-frequency noise) and epidemiological surveys.

The Commission would establish a Committee of national experts (without prejudice to the existence of Committees of experts already working on specific topics) which would assist it in drawing up the proposal laying down the general framework for the measures and the specific proposals mentioned above. This Committee would assist the Commission in its task of comparing national noise abatement programmes.

In its proposals for the Programme for 1982–86 the Commission proposed that in view of the Community's economic situation, future noise-abatement measures, while still aimed at the promotion of quieter products, would give much greater consideration to their socio-economic consequences.

Particular attention would be paid to the connection between noise reduction and possible energy savings.

Further, to give more consideration to the environmental impact of different types of noise and to avoid devising solutions that are too partial, a greater effort would be made:

- to determine simple physical indicators with a view to evaluating the quality of a particular acoustic environment;
- to explore the links between these indicators and the reactions of the populations subjected to different types of noise source, whether isolated or combined;
- to harmonize methods of forecasting levels of exposure to noise.

Standards for combining sound and heat insulation properties would be studied.

Particular interest would be paid to the problems of mechanical vibrations, especially as regards the protection of the Community's cultural heritage.

More generally, the Commission would seek to speed up the work being done on standardization by various competent international bodies.

7

Economic Aspects of Pollution Control

1. POLLUTER PAYS PRINCIPLE

One of the general principles endorsed by the Council, when it adopted the Environment Programme of the European Communities, was that the 'polluter should pay'. The Council recognized that the cost of preventing and eliminating nuisances must in principle be borne by the polluter. However, there might be certain exceptions and special arrangements, in particular for transitional periods, provided that they caused no significant distortion to international trade and investment. The Council said that, without prejudice to the application of the provisions of the Treaties, this principle should be stated explicitly and the arrangements for its application including the exceptions thereto should be defined at Community level. Where exceptions were made, the Council added that the need to progressively eliminate regional imbalances in the Community should also be taken into account.

On 5 March 1974 the Commission submitted to the Council a draft communication regarding cost allocation and actions by public authorities on environmental matters.

On 3 March 1975 the Council adopted a Recommendation which called upon Member States to conform in respect of allocation of costs and of action by public authorities in the field of environmental protection to the principles and rules governing their application which were contained in the Commission communication.

These principles and detailed rules are set out in an Annex to the Council's recommendation of 3 March 1975.

The Commission, in its communication to the Council, defines a polluter as someone who directly or indirectly damages the environment or creates conditions leading to such damage.

The Communication states that, depending on the instruments used and without prejudice to any compensation due under international law or national law, and/or regulations to be drawn up within the Community, polluters will be obliged to bear:

- expenditure on pollution control measures (investment in anti-pollution installations and equipment, introduction of new processes, cost of running anti-pollution installations, etc.) even when these go beyond the standards laid down by the public authorities.
- the charges.

The Communication defines the purpose of charges as being to encourage the polluter to take the necessary measures to reduce the pollution he is causing as cheaply as possible (incentive function) and/or to make him pay his share of the costs of collective measures, for example purification costs (redistribution function). The charges should be applied, according to the extent of pollution emitted, on the basis of an appropriate administrative procedure.

According to the Commission, charges should be fixed so that primarily they fulfil their incentive function.

In so far as the main function of charges is redistribution, they should at least be fixed within the context of the above-mentioned measures so that, for a given region and/or qualitative objective, the aggregate amount of the charges is equal to the total cost to the community of eliminating nuisances.

Income from charges may be used to finance either measures taken by public authorities or to help finance installations set up by an individual polluter, provided that the latter, at the specific request of the public authorities, is seen to render a particular service to the community, by reducing his pollution level to below that set by the competent authorities. In the latter instance, the financial aid granted must be limited to compensating for the services thus rendered by the polluter to the community.

In line with Article 92 et seq. of the EEC Treaty, income from charges may also be used to finance the installations of individual polluters for protecting the environment, in order actively to reduce existing pollution. In this case, the measures for financing should be incorporated in an official multi-annual finance programme by the competent authorities.

Where the overall revenue exceeds the total expenditure by the public authorities when applying the two preceding paragraphs, the surplus should preferably be used by each government for its national environment policies:however, the surplus may be used for granting aid only under the specified conditions.

In fact the Communication provides that aid may be granted for a limited period and possibly of a degressive nature, where the immediate application of very stringent standards or the imposition of substantial charges is likely to lead to serious economic disturbances and the rapid incorporation of pollution control costs into production costs may give rise to greater social costs.

The Communication recognizes that it may also prove necessary to allow some polluters time to adapt their products or production processes to the new standards. In any case, the communication states that such measures may apply only to existing production plants and existing products. (The enlargement or the transfer of existing production plants is to be considered as the creation of new plants where this represents an increase in productive capacity.)

The Commission's communication, annexed to the Council's recommendation of 3 March 1975, also specified that exceptions to the 'polluter pays' principle may be justified where, in the context of other policies (e.g. regional, industrial, social, and agricultural policies or scientific research and development policy), investment affecting environmental protection benefits from aid intended to solve certain industrial, agricultural or regional structural problems.

The Commission's communication lays down that the following shall not be considered contrary to the 'polluter pays' principle:

● financial contributions which might be granted to local authorities for the construction and operation of public installations for the protection of the environment, the cost of which could not be wholly covered in the short term from the charges paid by polluters using them. In so far as other effluent as well as household waste is treated in these installations, the service thus rendered to undertakings should be charged to them on the basis of the actual cost of the treatment concerned;
● financing designed to compensate for the particularly heavy costs which some polluters would be obliged to meet in order to achieve an exceptional degree of environmental cleanliness;
● contributions granted to foster activities concerning research and development with a view to implementing techniques, manufacturing processes and products causing less pollution.

2. COMMISSION MEMORANDUM ON STATE AIDS

In a separate memorandum to the Member States dated 6 November 1974 regarding the Community approach to state aids in environmental matters, the Commission itself expressed the view that during a transitional period state aids designed to assist existing firms in adapting to laws or regulations imposing major new burdens relating to environmental protection would qualify for exemption under Article 92(3)(b) EEC by being aids to promote the execution of an important project of common European interest.

This would apply only for the six-year period from 1 January 1975 to 31 December 1980. The Commission calculated that this should be long enough to enable all the Member States to implement arrangements ensuring that the polluter pays principle was applied throughout the Community on broadly similar principles.

In order to qualify for exemption under Article 92(3)(b), national aids would have to satisfy the following tests:

- they would have to be necessitated by new major obligations imposed by the State or by the Community on the recipient firms in relation to environmental protection.
- they would have to be granted to finance investments necessary to the adaptation which these firms would have to make to their plants in operation at 1 January 1975 in order to satisfy the stated obligations.

Such additional investment might be involved either in acquiring new equipment to reduce or eliminate pollution or nuisances or in adopting new production processes having the same effect; in the latter case aid should not be granted in respect to that part of the new investment whose effect is to increase productive capacity. The cost of replacing and operating the investments should be fully borne by the relevant firms.

When expressed as a net after-tax subsidy calculated by reference to the common method set out in the Commission Memorandum to the Council on Regional Aid Schemes, aids must not exceed:

- 45% for investments in 1975 and 1976;
- 30% for investments in 1977 and 1978;
- 15% for investments in 1979 and 1980.

The Commission felt that this degressive scale was justified because the Member States must be aware of the need to make polluters pay the price of their pollution as quickly as possible and because firms must be made to treat the investments required to eliminate pollution as a matter of urgency.

The maximum aid, although it was high, took account of the degree of effort required of businesses which had thought out their activities in an economic context where environmental costs were insufficiently taken into account, while the fact that it was always less than 50% accentuated the fact that, although not immediately applicable to its fully extent, the polluter pays principle remained the objective.

It went without saying, moreover, that these maxima would also have to be respected where, in a given Member State, the relevant investments might benefit from several specifically environmental aid schemes, at once.

Each year the Member States would have to give a statistical report for the past year on the aids granted and the investments involved in each industry, expressed as net subsidies.

As required by Article 93(1), the Commission would thus be in a position to monitor the application of these aid schemes and to act where necessary in order to tighten discipline should it be found that the schemes were liable to create problems in certain industries as regards competition and trade within the Community.

Within these limits the Member States would be able to implement both aid schemes in favour of given industries or regions and general schemes applicable to any particular industry or region. Any scheme which does not meet the above conditions would have to be modified, for otherwise the Commission would have to declare it incompatible with the common market.

The Memorandum from the Commission to the Member States also lays down certain guidelines for dealing during the transitional period with aids not satisfying the stated criteria, and after the transitional period with all specifically environmental aids.

A second memorandum from the Commission to member states was sent on 8 July 1980. The Commission took the view that the transitional period was too short to achieve the objectives set. The reasons were:

(a) the economic recession which set in at the beginning of the transitional period coupled to the need for industry to adjust to the new international situation, meant that the funds member states were able to set aside for environmental protection had to be restricted and attempts to regulate the question were hampered;

(b) the problems relating to the protection of the environment and preparation of the relevant laws and regulations was therefore a long and difficult task. Even more time was required to implement these provisions in cases where they had to be drawn up at Community level for subsequent national application;

(c) delays caused by the above developments have attracted particular attention because of increasing public demands for actions and improvements to the environment extending far beyond what so far has been achieved by governments.

The Commission therefore extended the transitional period to 31 December 1986. The rate of aid should not exceed 15% of the value of the investment aided. Only undertakings having installations in operation for at least two years before entry into force of the standards in question may qualify for assistance.

3. METHODOLOGY FOR ASSESSING POLLUTION CONTROL COSTS

On 8 December 1977, the Commission sent to the Council a draft Council recommendation to the Member States regarding methods of evaluating the cost of pollution control to industry. The Commission noted, in sending this draft Recommendation, that different methods of evaluating actual or probable pollution control costs were still used in different Member States. The data obtained were therefore seldom directly comparable at Community level. The Commission believed that it was necessary to adopt a common set of rules to which all future studies of pollution control costs in industry conducted in the Member State should conform.

Possible methods had been tried in practical sectoral studies within the Commission, by Member States themselves and on behalf of the OECD Environment Committee. In the light of these studies and of extensive discussion in the Commission's Group of Environmental Economic Experts of the problems they had raised, the Commission believed that it was opportune to propose a single methodology for future pollution control cost studies of particular sectors of industry within the Community which would ensure a minimum of comparability of the results they produce. The Council adopted the draft recommendation on 19 December 1978.

4. FUTURE WORK

In adopting the second Environment Programme, the Council laid increased stress on the economic aspects of pollution control. The Council recalled that the 1973 action programme included the statements that 'the protection of the environment against pollution and its improvement by the taking into consideration the quality of life in the decision-making machinery and production structures inevitably involves various kinds of expenditure', and that 'it is essential that the authorities make accurate assessments of the size of this expenditure in order to have a clear idea of what the economic, financial and social repercussions of proposed decisions are likely to be, and to adapt accordingly the procedures for implementing these decisions'.

It noted that, while the cost of measures to protect and improve the environment may be fairly limited at the macroeconomic level, they might be high at a sectoral level, i.e. for a particular industry or firm, a public body, local authority or for private individuals. The Commission should take account of the impact at microeconomic level of the various measures contemplated by costing the proposed measures, and, where relevant their effect on the prices of the products concerned, taking into account the required objectives and assessing the results of anti-pollution measures and the capacity of the firms, public bodies, local authorities or private individuals concerned to bear these costs.

The Commission should also consider the effects which the proposed measures might have on international competition, development and employment.

The Council went on to note, however, that when assessing the advantages of the proposals, comparison with the costs would not always be possible without attributing some particular interpretation and weighting to the advantages. First, improvement of the quality of the environment, which represents the beneficial effects of the measures taken, often cannot be assessed in monetary terms. This means that in such cases it is impossible to compare the benefits directly with the costs involved in implementing the measures; however, in these cases some cruder measure of output or benefit in more physical terms will often be possible and very valuable. Secondly, it is to be expected that the implementation of environmental measures will generally encourage industry to perfect less expensive anti-pollution techniques. This means that the cost of anti-pollution measures— which is measured on the basis of the state of advancement of the technique—will usually be over-estimated as compared with long-term costs.

The Council called on the Commission to continue the methodological and statistical work which it had begun under the 1973 action programme and to study the effectiveness of various economic instruments. It also asked the Commission in collaboration with a group of economic experts to give further thought to the strict application of the 'polluter pays principle'.

As already noted in the section on the Third Environment Programme (Chapter 1 section 7) environment policy is seen as a structural policy to be carried without regard to short term fluctuations in cyclical conditions in order to prevent natural resources from being seriously despoiled and to ensure future development potential is not sacrificed. Future action under the Commission proposals for 1982–1986 can be considered under two headings.

(a) Socio-economic context of the 1980s

Consolidation and the continuation of the measures laid down in the 1973 and 1977 programmes must take into account the socio-economic context of the 1980s and the new political and geographical dimensions of the Community of Ten; nor must pre-accession negotiations in progress with two other states be forgotten. Accordingly, the environment policy tries to link up with, and support, several major objectives for the whole Community economy over the next few years, especially that of economic recovery.

The socio-economic context of the 1980s will mean that environmental action must not only take account of the major problems confronting the Community (employment, inflation, energy, balance of payments and growing regional disparities) but must also contribute to the efforts made in other ways to find solutions.

Accordingly, environment policy must be concerned:

● to help in creating new jobs by the promotion and stimulation of the development of key industries with regard to products, equipment and processes that are either less polluting or use fewer non-renewable resources;
● to reduce any form of pollution or nuisance, or of interference with spatial features, the environment or resources which creates waste or unacceptable cost for the Community;

- to economize certain raw materials that are non-renewable, or of which supplies can be obtained only with difficulty, and to encourage the recycling of waste and the search for less polluting alternatives;
- to prevent or reduce the possible negative effects of using energy resources other than oil, such as coal or nuclear power, and to promote energy saving and the use of less polluting energy resources;
- to reinforce the implementation of the Information Agreement of 5 March 1973 to avoid individual national measures affecting the proper functioning of the internal market or making the adoption of appropriate Community measures more difficult.

(b) Creation of an overall strategy

One of two crucial principles if the goal of the creation of an overall strategy is to be achieved is that prevention rather than cure is the rule. Amongst the conditions which must be met if the principle of prevention is to have full effect is that an effort must be made to achieve optimum distribution of resources. In this respect the Commission states that:

1. Care must be taken to ensure optimum use of resources in a period of general economic difficulty. In the circumstances, the reasons for any new measures need to be examined carefully and their cost-effectiveness analysed before the measures are actually taken. The Commission will do everything necessary to strengthen this practice, which it has been applying in principle since it began its action.
2. The application of the polluter-pays principle is of decisive importance in a strategy which is designed to make the best use of resources. Apportioning the costs of protecting the environment to polluters, as provided by this principle, constitutes an incentive to them to reduce pollution caused by their activities and to discover less polluting products or technologies. This principle is therefore the chief way of bringing market forces to bear so as to achieve optimum structuring within a market economy.
3. The polluter-pays principle is usually applied by subjecting polluters to standards and/or charges and it implies that, in conformity with the general principles of the EEC Treaty concerning State aids that public authorities do not finance investments required to reduce pollution. However, in some cases the introduction of new obligations for this purpose could be delayed because the consequential financial burden can cause serious difficulties for older firms and thus for employment. Faced with this situation, the Commission, by two Decisions dating from 1974 and 1980, has accepted that Member States may grant aids, during a period ending in 1987 and under certain conditions, aimed at easing the introduction of new Regulations that could ensure adequate protection of the environment.

4.   A particular area where State aids may be needed is in the protection of nature and of the landscape, aids which are usually given to local authorities or to voluntary organisations. Even if these aids do not contravene the provisions of the EEC Treaty concerning State aids, it is desirable to bring some of these aids into a Community context and so ensure their cohesion and thus increase their effectiveness.

5.   Charges constitute one of the instruments for the application of the polluter-pays principle and they can provide an incentive to the introduction of anti-pollution measures to reinforce the application of standards and stimulate innovation, especially if residual pollution is also covered by the charge. It is necessary therefore to study carefully the fields where charging systems would allow the achievement of the objectives of environment policy more efficiently.

6.   An environmental strategy at Community level should also be funded by financial resources specifically set aside for environmental purposes. Such instruments, which by their very existence would have a snowball effect and tend to stimulate the required 'osmosis' between national and Community environmental policies, would put the Community in a better position to help implement a balanced environmental policy in all its regions. In this connection the European Parliament has proposed the creation of an EEC Environment Fund.

The Commission has proposed the inclusion in its preliminary draft budget for 1982 of (admittedly symbolic) amounts for these purposes. An optimum form for Community financial intervention concerning the environment will have to be thought out on the basis of the experience which would be gained from using these appropriations. (See Section 9.2.)

7.   There is a need to deploy greater efforts on integrating environmental data more fully into national accounts. National accounts in their traditional form do not take account of most environmental costs nor the benefits of improving the environment because they are difficult to measure. Consequently it is important to improve indicators of environmental quality to supplement traditional national accounts that take inadequate account of the costs and benefits of improving the environment. Moreover, the inclusion of environment costs in GNP would be facilitated by the application of the 'polluter-pays' principle and the use of charges to internalize external costs.

8

Research

1. DEVELOPMENT OF ENVIRONMENTAL RESEARCH ACTIVITIES OF THE EUROPEAN COMMUNITIES

Attempts to coordinate environmental research at Community level started in 1967* when the PREST (Working Party on Scientific and Technical Research Policy of the Medium-Term Economic Policy Committee) considering that environmental protection was a most appropriate subject for scientific cooperation in the Community, established an expert group on pollution and nuisances which, later on, was taken over by the COST (Scientific and Technical Co-operation) Committee and assigned to it the task of preparing research projects to be undertaken jointly. The first concrete results came out on 23 November 1971, with the signature of three COST agreements involving most Member States of the EEC as well as several third countries. These dealt with very specific topics, to wit the physico-chemical behaviour of sulphur dioxide in the atmosphere (COST 61a), the analysis of organic micropollutants in water (COST 64b) and sewage sludge processing (COST 68).

In the meantime preparations were made for the inclusion of non-nuclear research, i.e. on environmental protection in the programme of the Joint Research Centre (JRC) at Ispra, Italy. An arrangement was eventually concluded in 1972 for carrying out environmental research at the Ispra establishment under a contract between the six Member States and the JRC for a period of one year. This made it possible to establish the main lines of the direct action programme in this field, including an effective involvement in the subject matter of two of the COST projects (61a and 64b).

During 1972 and at the beginning of 1973 negotiations were held in parallel for the acceptance by the Council, on the one hand, of the Programme of Action on the Enviroment and, on the other, of a first Multi-

---

* It should be noted, however, that research on the effects of ionizing radiation and environmental implications thereof has been carried out under the Euratom treaty since 1961: the latest programme for the period 1980–84 was adopted on 13 March 1980.

annual Research Programme in environmental protection; the latter was considered from the outset as having to include both direct action to be carried out at the JRC, and indirect action involving a number of specialized organizations of the Member States. Indeed, the input from the JRC, albeit important in certain fields, had to be relatively limited in quantity in view of the skilled scientific manpower available. Yet it was felt that a Community programme, in order to make a significant impact in solving environmental problems, must reach a minimum critical size and form a coherent body of research bearing on a variety of subjects. Decisions were eventually made by the Council on 14 May and 18 June 1973 for the research programme and on 19 July and 22 November 1973 for the Action Programme.

The first common research programme (direct and indirect actions) was designed specifically to provide scientific and technical support to the sectoral policy of the Community on environment, essentially with regard to Part II, Title I 'Reduction of Pollution and Nuisances' of the Action Programme, and particularly to the actions concerned with (1) the objective evaluation of the risks to human health and to the environment from pollution, (2) the improvement of measurements of pollution, (3) the management of environmental information.

The direct action to which a total of 15.85 million units of account was allocated, includes research work under the following headings:

● analysis and monitoring of pollution (including development of a multi-detection unit, remote sensing of air pollutants, and a pilot data bank of environmental chemicals);
● fate and effects of pollutants;
● models and system analysis applied to the eutrophication of a lake and to air pollution
● theoretical studies on thermal pollution and catalytic oxidation of water pollutants.

The Council decision of 26 August 1975 which amended earlier decisions relating to the direct action research work of the European Economic Community laid down that the programme of direct research should be subject to review at the beginning of 1976.

The indirect action, to which a maximum amount of 6.3m u.a. was allocated until 31 December 1975 for the conclusion of shared-cost research contracts, comprises the following 6 topics:

(a) epidemiological surveys of the effects of air and water pollution;
(b) harmful effects of lead;
(c) effects of micropollutants on man;
(d) ecological effects of water pollution;
(e) remote sensing of air pollution;
(f) data bank on environmental chemicals.

2. IMPLEMENTATION OF THE FIRST ENVIRONMENTAL RESEARCH PROGRAMME (INDIRECT ACTION)

The first programme, decided by the Council on 18 June 1973, became actually operational with the creation, on 10 December 1973, of the Advisory Committee on Programme Management for Environmental Research which met for the first time on 28 and 29 January 1974. Prior to this, a call for research proposals published in the Official Journal of the European Communities on 28 July 1973, as well as contacts with governmental authorities involved in environmental research management in the Member States, resulted in the submission of a large number of applications.

The Commission services and the Advisory Committee examined these proposals and selected a number of them on the basis of the following criteria:

(a)   relevance to the overall programme;
(b)   scientific value of proposal;
(c)   possibility of coordination with other projects in both the direct and indirect actions as well as with national research programmes;
(d)   prospect of success of planned work;
(e)   cost;
(f)   provisions made for complementary finance.

In 1978, the Commission published a volume of final reports on research sponsored under the indirect action programme. In total, 127 shared-cost contracts (for which the maximum contribution of the Community was fixed as a general rule at 50% of the total cost) were signed, of which 21 were under topic 1, 24 under topic 2, 23 under topic 3, 35 under topic 4, 14 under topic 5, 10 under topic 6.

Following is a sample of the coordinated products initiated during the first programme:

(a)   an epidemiological survey aimed at establishing correlations between air pollution and respiratory diseases in schoolchildren, carried out simultaneously by 10 institutes following the same protocol and involving 20,000 subjects;
(b)   a study of pollution in streams of the Luxembourg-Saarland-Lorraine region, undertaken jointly by Belgian, French, German and Luxembourger laboratories in relation with the establishment of quality objectives for these watercourses;
(c)   a pilot data bank on environmental chemicals (ECDIN project) carried out by the JRC and seven other organizations; such a data bank makes it possible to collect, store and retrieve all relevant information needed, i.e. to prepare regulations on environmental chemicals and

to determine the best countermeasures in case of accidental contamination;

(d) a project for the development of mutagenicity testing methods for environmental pollutants, involving 9 laboratories, in order to improve the techniques for assessing long-term effects of pollution on human health;

(e) work by a group of laboratories on the development of remote sensing systems for atmospherical pollution by the use of lasers and other optical methods; they participated in a field campaign in July 1975 to compare the performance of their equipment;

(f) numerous researches on chronic toxicity of lead at low level have been started; they have given certain results which were used for the preparation of a draft directive of the Council on the monitoring of the degree of contamination of the population by lead; one project under way should make it possible to determine effectively the importance of lead from automobile exhausts in the total input of lead by populations;

(g) a project on the characterization of sewage sludges from effluent treatment plants has been completed which should facilitate the development of new techniques for the treatment and utilization of these sludges;

(h) a critical evaluation of the performance of two plants for the joint incineration of refuse and sewage sludges has been carried out;

(i) over 1,000 organic micropollutants have been identified in surface waters in view of evaluating the potential toxicity of these waters and to guide the development of treatments for drinking water.

In order to assist the Commission Services and the Advisory Committee in the management of the programme, steering committees for certain closely related projects (e.g. epidemiological surveys) and contact groups in various fields were established to ensure the progressive coordination and complementarity of the work sponsored.

Three management committees were set up, one for each of the COST projects mentioned before, which are run as concerted actions. These entailed yearly expenditure of 1.3m u.a. for the Member States to which are added the contributions of participating third countries.

3. ENVIRONMENTAL RESEARCH PROGRAMME 1976–1980

The Council, on 15 March 1976, adopted a new pluriannual environmental research and development programme (indirect action) for a total amount of 16m u.a. for the period 1976 to 1980, which was centred on four main fields:

- Research aimed at the establishment of criteria (exposure-effect relationships): heavy metals, organic micropollutants, fibrous materials, new chemical products, air and water (fresh and sea) pollutants, thermal discharges and noise nuisances.
- R & D on environmental management and information; the effort was centred on the problems of new chemicals likely to hazard health or the environment. The ECDIN pilot project was extended and the findings analysed.
- R & D on the prevention and reduction of pollution and nuisances: special attention was given to treatment of waste water, sewage sludge, industrial effluent and waste processing.
- Improvement of the environment: the planned research concerned the structure and function of ecosystems, biogeochemical cycles, reclamation of spoiled or waste land, remote sensing of environmental disturbances and the ecological implications of land development and modern methods of farming.

On 30 June 1978, the Commission, following a review of the existing environmental R & D programme by the Advisory Committee on Programme Management for Environmental Research, sent to the Council a proposal to modify this 5-year programme in order to step up research into the pollution problems it saw as posing the greatest threat to human health and the physical environment. A revised programme was adopted on 9 October 1979.

Area 1 (research aimed at the establishment of criteria i.e. dose-effect relations for pollutants and environmental chemicals)

Area 2 (R & D on environmental information management, concerning essentially environmental chemicals
This area involves only 4% of total funds and is mainly concerned with the collection and handling of data on chemicals in the framework of the ECDIN (Environmental Chemicals Data and Information Network) project. The EEC's current research effort in this field will be maintained. The Commission notes that the Community will require a system to store and retrieve data submitted in the notification procedure for new chemicals provided for in the proposed Directive covering marketing procedures for chemicals then under discussion in the Council. Other Directives and draft Directives, such as those pertaining to toxic and dangerous waste and the control of certain industrial activities (in preparation) will also necessitate the collection and handling of large amounts of data on chemicals.

Area 3 (R & D on the reduction and prevention of pollution and nuisances)
Approximately 10% of programme funds—a substantial increase—will be devoted to research into advanced biological and physico-chemical treatment of waste water. The Commission points out that in view of the

implementation of EEC water quality Directives there is a need to test the performance of advanced water treatment methods and the relevance of existing parameters to test the effectiveness of treatment. In addition, the Commission considers that a review of national R & D activities in this field should be made to ascertain whether a concerted action should be launched, taking into account existing international efforts.

Area 4 (R & D concerning the protection and improvement of the natural environment)
With 15–20% of total research funds this area will cover the following topics:

- Ecosystem ecology and biogeochemical cycles, especially studies of ecosystems (contribution to the establishment of an ecological cartography); ozone shield depletion in the stratosphere and CO_2 accumulation in the atmosphere;
- reclamation of derelict land;
- bird protection: population dynamics and habitat protection.

The total cost to the EEC budget of the revised programme is put at 20.8 million European units of account—an increase of 4.8 e.u.a. on the original programme allocation of 16 million units of account.

4. ENVIRONMENTAL RESEARCH PROGRAMME 1981–85

On 30 June 1980 the Commission submitted to the Council a proposal for a decision adopting a sectoral research and development programme (environmental protection and climatology) (indirect and concerted action) 1981–85. The Council at its 619th meeting on 20 December 1979 made certain recommendations on the Community's research policy generally. The Council recognized the environment as a priority interest and recommended the concentration of Community R & D programmes and the examination of the possibility of setting Community indirect and concerted action programmes in the context of an appropriate multiannual framework programme while rationalizing structures for the preparation, examination and implementation of Community R & D programmes.

In line with these recommendations the Commission prepared the proposal in which an attempt was made to group several concerted and indirect research actions pertaining broadly to the field of environment.

The following R & D actions were on hand at the end of 1979 (in addition to the direct action):

- second R & D programme in the field of environment (indirect action) (1976–1980), revised on 9 October 1979;

- Community concerted action on the treatment and utilization of sewage sludge and Community-COST concertation agreement (COST 68bis), (27 September 1977–26 September 1980);
- Community concerted action on the analysis of organic micropollutants in water and Community-COST concertation agreement (COST 64b bis) (November 1978–October 1982);
- Community concerted action on the physico-chemical behaviour of atmospheric pollutants, and Community-COST concertation agreement (COST 61a bis) (November 1978–October 1982);
- COST action on marine benthic ecosystems with Community participation (COST 47) (April 1979–April 1984);
- research programme in the field of climatology (indirect action) (1980–1984);

The proposal grouped these indirect and concerted actions into two sub-programmes: Environment protection and Climatology. The decision was adopted on 3 March 1981. It has three main aims:

(a) to provide scientific and technical data to support the Community Environment Action Programme;
(b) to address less immediate problems and prepare the way for policies to be enacted in the medium to long term taking care of foreseeable environmental trends;
(c) to further the co-ordination at Community level of national research activities in the environmental field.

It provides for:

Sub-programme I: Environment Protection

Research area 1: Sources, pathways and effects of pollutants: Heavy metals, organic micro-pollutants and new chemicals (including COST 64b bis), asbestos and other fibres, air quality (including COST 61a bis), surface and underground freshwater quality, thermal pollution, marine environmental quality (including COST 47) and noise pollution.
Research area 2: Reduction and prevention of pollution and nuisances: Sewage sludge (including COST 68 bis), pollution abatement technologies, clean technologies, ecological effects of solid waste disposal, oil pollution cleaning techniques, impact of new technologies.
Research area 3: Protection, conservation and management of natural environments: Ecosystems studies, biogeochemical cycles, ecosystems conservation, bird protection, reclamation of damaged ecosystems.
Research area 4: Environment information management: Data bank on environmental chemicals, evaluation, storage and exploration of data, ecological cartography.
Research area 5: Complex interactive systems; man–environment interactions

Sub-programme II: Climatology (indirect action)

Research area 1: Understanding climate: Reconstruction of past climate, climate modelling and prediction.
Research area 2: Man–climate interactions: Climate variability and European resources, man's influence on climate.
Service activities: interdisciplinary studies for analysis, evaluation and application of the results of Research area 2, inventory, co-ordination and enrichment of European climatic data sets.

Funds will be distributed as follows:

Sub-programme I: 33m ecu: indirect action
                1m ecu: concerted action
   Research area 1: 50 to 55%
   Research area 2: 20 to 25%
   Research area 3: 15 to 20%
   Research area 4: 5 to 10%
   Research area 5: 1 to 5%
Sub-programme II: 8m ecu.

Execution of the programme for indirect action and coordination of concerted actions will be the responsibility of the Commission assisted by the Advisory Committee on Programme Management in the field of Environment Protection and in the field of Climatology.

The programme will be carried out in close coordination with other research programmes of the EC which include an environmental component. These are:

- the parts of the Joint Research Centre programme (1980–1983) 'Protection of the Environment' and 'Remote Sensing from Space' in particular the following projects:
  - ECDIN
  - exposure to chemicals, particularly indoor pollution and organic compounds
  - air quality
  - water quality
  - heavy metals and health effects
  - environmental impact of conventional power plants
  - remote sensing: protection of the sea
- the R & D programme on recycling of urban and industrial waste (1979–82)
- the R & D programme on recycling of paper and board (1978–1981)
- the Agricultural research programme (1979–1983) in particular its Part A: Socio-structural objectives
  - land use and rural resources

● effluents and wastes from agriculture
● the European Steel and Coal Community research projects in particular those concerned with environment problems in the steel industry.

Problems of radioactive contamination of the environment are addressed in two research programmes under the Euratom treaties:

● Biology-health protection (1980–84)
● Radioactive waste management and storage (1980–84) (see page 131)

A subcommittee to deal with environment research was set up in 1982 by the Community's Scientific and Technical Research Committee (CREST). Its tasks are to access the extent of research in environmental matters undertaken by Member States both on a national and European level.

5. RECYCLING OF URBAN AND INDUSTRIAL WASTE (1979–82)

A research and development programme on recycling of urban and industrail waste was authorized on 20 November 1979 under a Council Decision. The second phase of this, for which 1.3m ecu has been allocated, covers three categories: technology for the sorting of bulk household waste; energy recovery; and development of mechanical-biological processing concepts for household waste with high organic content.

9

Preventive Action

1. ENVIRONMENTAL ASSESSMENT

In the introduction to the proposed Action Programme for 1982 to 1986, the Commission points out that initially the essential aim of the Community environment policy was the control of pollution and nuisances. It has, however, gradually assumed an overall preventive character. This development is seen as offering the twofold advantage of environmental protection measures at the lowest possible cost and of positive measures which support and complement economic development. The establishment in the Community of a system of prior inspection of new chemical substances (see Chapter 5 Section 2) was the first milestone in a wider-ranging policy; the second was the setting up in the Community of a system for taking the measures needed to prevent accidents and to limit the consequences of accidents arising from dangerous substances in industrial activities particularly from certain more dangerous substances.

A third major step would be the adoption of the proposal put forward by the Commission for a directive on the assessment of the environmental effects of certain public and private effects.

This proposal was submitted to the Council on 16 June 1980. As a result of the Parliament's recommendations certain amendments were proposed by the Commission in its letter to the Council of 5 April 1982. The proposal, as amended, provides for mandatory prior assessment of environmental effects in respect of new projects which appear likely to have significant effects on the environment. Three groups of projects are identified: the first group (see Table 9.1) includes projects that by reason of their size and/or the amount of pollution they cause are likely significantly to affect the environment under any circumstances. A second class of project (see Table 9.2) consists of those likely to produce significant effects only under certain conditions e.g. when they reach a certain size or produce a specific amount of pollution; the third group consists of projects which are unlikely to produce significant effects on the environment. For projects in the first group a full assessment would have to be carried out; for those in the second group a simplified assessment would be carried out; and those

Table 9.1: Commission proposal for directive of 16 June 1980 as amended by proposal of 5 April 1982. Assessment of the environmental effects of certain public and private projects

Development projects (1) for which full assessment would be required
1. *Extractive industry* Extraction and briquetting of solid fuels (11) Extraction of bituminous shale (133) Extraction of ores containing fissionable and fertile material (151) Extraction and preparation of metalliferous ores (21)
2. *Energy industry* Coke ovens (12) Petroleum refining (140.1) Production and processing of fissionable and fertile materials (152) Generation of electricity from nuclear energy (161.3) Coal gasification plants Disposal facilities for radioactive waste
3. *Production and preliminary processing of metals* (22) Iron and steel industry, excluding integrated coke ovens (221) Cold rolling of steel (223) Production and primary processing of non-ferrous metals and ferro-alloys (224)
4. *Manufacture of non-metallic mineral products* (24) Manufacture of cement (242.1) Manufacture of asbestos-cement products (243.1) Manufacture of blue asbestos
5. *Chemical industry* (25) Petrochemical complexes for the production of olefins, olefine derivatives, bulk monomers and polymers Chemical complexes for the production of organic basic intermediates Complexes for the production of basic inorganic chemicals
6. *Metal manufacture* (3) Foundries (311) Forging (312.11) Treatment and coating of metals (313.5) Manufacture of aeroplane and helicopter engines (364.1)
7. *Food industry* (41/42) Slaughter-houses (412.1) Manufacture and refining of sugar (420.1, 420.2) Manufacture of starch and starch products (418)
8. *Processing of rubber* (48) Factories for the primary production of rubber Manufacture of rubber tyres (481.1) Factories for the renewal or reprocessing of rubber production
9. *Building and civil engineering* (50) Construction of motorways Intercity railways, including high speed tracks Airports Commercial harbours Construction of waterways for inland navigation

Table 9.1 continued

Permanent motor and motorcycle racing tracks Installation of surface pipelines for long distance transport

(1) Development projects are classified, as far as possible, in the classes, groups and sub-groups of the 'General Industrial Classification of Economic Activities' within the European Community adopted by the Statistical Office of the European Communities, 1970. Reference numbers of the classification are indicated, where applicable.

Table 9.2: Commission proposal for draft directive of 16 June 1980 as amended by proposal of 5 April 1982. Assessment of environmental effects of certain public and private projects

Projects (1) for which a simplified assessment would be required
1. *Agriculture* Projects of land reform Projects for cultivating natural areas and abandoned land Water management projects for agriculture (drainage, irrigation) Intensive livestock rearing units Major changes in management plans for important forest areas
2. *Extractive industry* Extraction of petroleum (131) Extraction and purifying of natural gas (132) Other deep drillings Extraction of minerals other than metalliferous and energy-producing minerals (23)
3. *Energy industry* Research plants for the production and processing of fissionable and fertile material Production and distribution of electricity, gas, steam and hot water (except the production of electricity from nuclear energy) (16) Storage of fossil fuels
4. *Production and preliminary processing of metals* Manufacture of steel tubes (222) Drawing and cold folding of steel (223)
5. *Manufacture of glass fibre (247.5) glass wool and silicate wool*
6. *Chemical industry* Production and treatment of intermediate products and fine chemicals Production of pesticides and pharmaceutical products, paint and varnishes, elastomers and paroxides Storage facilities for petroleum, petrochemical and chemical products
7. *Metal manufacture (3)* Stamping, pressing (312.2) Secondary transformation treatment and coating of metals (313) Boilermaking, manufacture of reservoirs, tanks and other sheet-metal containers (315) Manufacture and assembly of motor vehicles (including road tractors) and manufacture of motor vehicle engines (351) Manufacture of other means of transport (36)
8. *Food industry (41/42)* manufacture and vegetable and animal oils and fats (411)

Table 9.2 continued

	Processing and conserving of meat (412.2) Manufacture of dairy products Brewing and malting (427) Fish-meal and fish-oil factories
9.	*Textile, leather, wood, paper industry* Wool washing and degreasing factories Tanning and dressing factories (441.1) Manufacture of veneer and plywood (462.1) Manufacture of fibre board and of particle board (462.2) Manufacture of pulp, paper and board (471) Cellulose mills Textile dyeworks
10.	*Building and civil engineering* (30) Major projects for industrial estates Major urban projects Major tourist installations Construction of roads, harbours, airfields River draining and flood relief works Hydroelectric and irrigation dams Impounding reservoirs Installations for the disposal of industrial and domestic waste Storage of scrap iron
11.	*Modification to development projects included in Annex I*

(1) The projects are classified, as far as possible, in the classes, groups and sub-groups of the 'General Industrial Classification of Economic Activities' within the European Community adopted by the Statistical Office of the European Communities, 1970. Reference numbers of the classification are indicated, where applicable.

in the third group would not as a general rule require an assessment. There is provision for Member States to exempt in exceptional circumstance projects in the first list from assessment. Competent authorities in Member States would determine whether projects in the third group should be made subject to an assessment and whether this should be in full or in simplified form.

The assessment of the effects on the environment would consist of a number of steps taken by the various parties concerned with the implementation of a project: the collection and supply of the relevant information on the likely effects on the environment by the developer of the project; the consultation of governmental departments and bodies (e.g. those responsible for pollution control) and of the public by the competent authorities; and the drawing up of an assessment document by the competent authorities.

The assessment would identify, describe and evaluate as appropriate the direct and indirect effects of a project on human beings, flora and fauna, soil, water, air, climatic factors, material assets, including the cultural heritage, and the landscapes, natural resources and the ecological balance.

The draft Directive provides that the developer shall provide along with the application for authorization, in particular:

- a description of the proposed project and, where applicable, of the reasonable alternatives for the site and/or design of the project;
- a description of the environmental features likely to be significantly affected by the proposed project, including where applicable, those located in another Member State;
- an assessment of the likely significant effects of the project on the environment, including, where applicable, effects on the environment of another Member State;
- a description of the measures envisaged to eliminate, reduce or compensate those impacts;
- review of the relationship between the proposed project and existing environmental and land-use plans and standards for the area likely to be affected;
- in the case of significant effects on the environment, an explanation of the reasons for the choice of the site and/or design of the proposed project, compared with reasonable alternative solutions having less effects on the environment, if any;
- a non-technical summary of the items above, to the extent that this information is relevant to the stage of the authorization and to the specific characteristics of the project and the environment likely to be affected.

The proposal provides for publishing the fact that an application for authorization of a project has been made, for making public the application and the information supplied by the developer and for making publicly available the competent authority's assessment, comments and opinions received, and reasons for granting or refusing authorization. There is also a requirement to respect limitations imposed by national legislation and practices with respect to industrial and commercial secrecy. Special provisions for consultation in cases where there could be effects across boundaries with other Member States are included.

2. AN EEC ENVIRONMENT FUND

On 11 January 1983, the Commission made a proposal for a Council Regulation on action by the Community relating to the environment (ACE). The Commission noted that substantial results had been achieved in various ways in a very short space of time. The main endeavours had been in the form of legislation aimed at reducing pollution and preserving the natural environment. In just under seven years the Community had adopted over 60 pieces of legislation in this field, including fifteen on the reduction of air pollution, seven on waste, eight on noise abatement, and four on the protection of the environment, land and natural resources. The Commission pointed out, however, that rules and regulations alone were an insufficient basis for a genuinely dynamic environmental protection policy; as such a policy must, above all, be preventive in nature, recourse should also be had to other methods and measures of support.

The European Parliament, convinced of the necessity to create for this purpose an 'EEC Environment Fund', had taken the initiative of entering in the 1982 budget an article (661) entitled 'Community operations concerning the environment'. The total budget allocation for the four new budget headings within that article amounted to 6.5 million ECU for 1982.

Taking into account the experience which it had acquired in using these budget headings for descriptive analysis and pilot experiments, the Commission considered that two of these headings should be used in such a way as to go beyond the stage of *ad hoc* measures, in order to make the best possible use of the financial resources available.

The headings in question concerned 'clean technologies' and the protection of certain sensitive areas of Community interest. These seemed to be two sectors concerning which such new forms of action were most desirable and urgent and for which no other existing financial instrument available to the Community seemed to be appropriate.

The Commission stated that the Community could in this fashion also make a contribution towards the practical utilization of the results of the Community's environmental research programmes and, where clean technologies were concerned, the results of the research programme in the raw materials sector.

(a) Development of new 'clean' technologies

The second Community action programme on the environment, which was adopted in 1977, had as one of its objectives the identification, in respect of each polluting branch of industry, of the technical or other processes likely to reduce, eliminate or prevent the emission of polluting substances or the creation of nuisances.

At its meetings on 18 December 1978 and 9 April 1979, the Council emphasized the role of clean technologies, specifying three aims for them:

(a) to cause less pollution, i.e. discharge less effluent into the natural environment,
(b) to produce less waste,
(c) to be more economical of natural resources (in particular raw materials).

It was, moreover, essential to improve still further the monitoring of the environment, and new techniques and methods were needed for this purpose.

In proposing the draft Regulation, the Commission stated that 'the objective would be to aid firms, federations of firms or other bodies answering an invitation, published by the Commission, for the submission of pilot projects intended to attain all or some of the three above-mentioned objectives or to improve methods and equipment, for monitoring the various sectors of the environment.

Applications should contain the following information:

- a detailed description of the project, and in particular the organization of its administration and the results expected,
- the time-scale for carrying out the project,
- the nature and extent of the technical and economic problems involved in the project;
- the cost of the project, its viability and the financing arrangements proposed;
- the extent to which the relevant experience may provide encouragement for the widespread introduction of the technique, process or product in the Community, the general application prospects for this technique, process or product and the benefits thereby obtainable for the economy as a whole;
- any other information which may justify the Community support requested,
- how it is proposed to disseminate the results of the project.

In order to be able to assess the value of the projects submitted for Community aid, the Commission would consult independent technical experts before referring the projects to an *ad hoc* Advisory Committee— the creation of which was proposed—for an opinion.

The Commission noted that it seemed to be particularly desirable that precedence should be given, in the granting of Community aid, to small and medium-sized firms whose financial resources were limited but whose innovative capacity in the field of clean technologies would seem to be considerable and could make a direct contribution towards reducing the social costs of pollution and nuisance in the Community.

It was proposed that the aid should be granted directly to natural or legal persons answering an invitation to submit projects published in the Official Journal of the European Communities. Such persons should undertake to inform the Commission, within a period to be determined contractually by mutual agreement, of progress made in attaining the stated objectives.

(b) Protection of the natural environment in certain sensitive areas of Community interest

Certain problems relating to the preservation of the natural environment were of Community interest and of such a scale as to necessitate a Community approach, but not solely in the form of rules and regulations. The 'preventive preservation' (i.e. their protection from pollution or undesirable forms of development etc.) of certain sensitive natural areas, in accordance with certain Community directives or international conventions to which the Community directives or international conventions to which the Community is, or is preparing to be, a Contracting Party, was of prime importance. These areas were geographically identifiable following studies carried out at the Commission's request.

The Commission pointed out that the protection of flora and fauna was an essential part of nature conservation. At Community level, the main piece of legislation approved by the Council was the Directive of 2 April 1979 on the conservation of wild birds. This Directive, and the Council resolution of 2 April 1979 relating to it, placed particular emphasis on the preservation of areas which provide a habitat for certain species of flora (biotopes). There were similar obligations arising from the Berne and Bonn Conventions and the Protocol to the Barcelona Convention, to which the Community is, or was preparing to be, a Contracting Party.

It was proposed that aid should be granted to public authorities or other bodies (including individuals) recognized by these authorities, for the conservation, management or acquisition of such areas by such authorities.

Applications should contain a detailed description of the project, and in particular the following information:

- the organization of its administration, and the expected results,
- the time-scale for carrying out the project,
- the nature and extent of the problems which the project is intended to resolve,
- the cost of the project, its viability and the financing arrangements proposed,
- any other information which may justify the application,
- how it is proposed to disseminate the results of the project.

The draft Regulation provided that financial compensation could also be granted where action was taken to restrict, transfer or put an end to activities incompatible with the use and status of these areas. None of the other existing financial instruments available to the Community could provide support for such operations as their objectives were fundamentally very different.

Community aid would take the form of financial intervention representing a percentage (not exceeding 50% for investments and financial compensation for the restriction of certain economic activities, and not exceeding 70% for descriptive analyses) of the cost of the operations.

The Commission would grant or refuse all or part of the aid requested under the Regulations after consulting an *ad hoc* Advisory Committee. This Committee would consist of representatives of the Member States and would be chaired by a Commission representative.

For 'clean' technologies, the Commission requested commitment appropriations of 3 million ECU for 1983, 4 million ECU for 1984 and 7 million

ECU for 1985 (corresponding figures for payment appropriations are 1.5 million, 2 million and 3.5 million ECU).

For 'protection of the natural environment', the Commission requested commitment appropriations of 3.5 million ECU for 1983, 4 million in 1984 and 6 million in 1985 (corresponding figures for payment appropriations are 1.8 million, 2 million and 3 million ECU).

The Commission stated that in 1982 it had received requests for financial intervention in excess of the funds available. This had made it necessary to adopt restrictive criteria for eligibility. The intention was to speed up the implementation of projects. The Community should participate financially in the pilot operations as well as studies, if the desired objective was to be achieved. The requirements for both types of appropriation for 1983 had been based on the large number of requests received to date.

The Commission's proposal for a Council Regulation is still being discussed by the Council. The European Parliament gave a favourite opinion, on the basis of a report by Mr Johnson, in April 1983.

10

Implementation of Directives

The implementation of the Directives is monitored in order to check that Member States are introducing the laws, regulations and administrative provisions necessary to comply within the given time-limits with the various Directives adopted by the Council.

To this end, the Commission examines the essential provisions of national law sent in by the Member States and where necessary initiates the procedure provided for in Article 169 of the EEC Treaty.

In May 1980 the Commission reported that, apart from the Directives on motor vehicles and other products which were adopted by the Council as part of the General Programme to remove technical barriers to trade in industrial products, eight environment Directives were to have been embodied in national law by 1 February 1980.

These included two Directives on water quality (surface water intended for the abstraction of drinking water and bathing water), four Directives on wastes (waste oils, wastes, PCBs, waste from the titanium dioxide industry) and two Directives on atmospheric pollution (the sulphur content of gas oils and the lead content of petrol).

After checking the implementation of these eight Directives, the Commission served thirty-three notices, published nineteen reasoned Opinions and brought one case before the Court of Justice.

Progress in implementing these Directives varied considerably from one Member State to another. Three Member States have complied completely or almost completely with the obligations imposed upon them by the provisions of these Directives. Four Member States have not enacted all of them, although they are in the process of doing so. Finally, two Member States (Belgium and Italy) had yet to begin implementation.

Since then a further 6 environment directives described in this publication (toxic and dangerous wastes, shellfish water, water for freshwater fish, noise from subsonic aircraft, sampling of surface water and ground water) should have been implemented. Up to the end of April 1982, the Commission in respect of these directives had served a further 14 notices, published 12 more reasoned opinions and brought 14 cases before the Court of Justice. Judgment has been delivered on 13 cases finding that the Member States concerned have failed to fulfil their obligations under the Treaty. In most cases steps have been taken by the Member States concerned to enact the necessary measures (see Table 10.1).

Table 10.1: Implementation of Directives: Steps taken by Commission and Court of Justice

March 80	Judgment v Italy		Sulphur in gas oil
May 80	Reasoned opinions	UK (2) I UX (2)	} Waste oils, PCB disposal, } Waste
		BE IT	} PCB disposal
June 80	Art 169 action	BE IT LU	} Lead in petrol
	Reasoned opinions	UK UK BE	Bathing water Surface water TiO₂ waste
Feb 81	Court of Justice	IT	5 actions (PCB disposal, waste oil, bathing water, surface water, waste)
Apr 81	Court of Justice	BE	6 actions (TiO₂ waste, waste oil, waste, PCB disposal, bathing water, surface water)
		NL	(Bathing water, surface water)
July 81	Reasoned opinions	LUX	Surface water
		BE IT	} Lead content of petrol
Oct 81	Art 169 action	BE NL Denmark IT	} Freshwater fish
		LUX IT Denmark	} Noise subsonic aircraft
Dec 81	Judgment v	IT	5 cases above
Feb 82	Art 169 action	LUX IT FR	Waste TiO₂ waste Noise subsonic aircraft
	Judgment v	BE	6 cases above
Apr 82	Judgment v Act 169 action Court of Justice	NL Ireland IT	Bathing water Noise subsonic aircraft Lead content of petrol

11

International Action

One of the objectives of Community environment policy is 'to seek common solutions to environment problems with States outside the Community particularly in international organizations'.

Furthermore the eighth principle of this policy states that 'the effectiveness of effort aimed at promoting global environmental research and policy will be increased by a clearly defined long-term concept of a European environmental policy.'

'In the spirit of the Declaration of the Heads of State or Government at Paris, the Community and the Member States must make their voices heard in the international organizations dealing with aspects of the environment and must make an original contribution in these organizations, with the authority which a common point of view confers on them.'

1. CONVENTION ON MARINE POLLUTION ARISING FROM LAND-BASED SOURCES (PARIS CONVENTION)

The first session of the Conference on Marine Pollution arising from Land-based Sources was held in Paris from 17–21 September 1973, at the invitation of the French Government. The following countries participated: Belgium, Denmark, France, Germany, Ireland, the Netherlands, Norway, Portugal, Spain, Sweden and the United Kingdom. Finland, Luxembourg, Iceland, Italy and Switzerland were present as observers. The Council of Europe, the United Nations Environment Programme (UNEP) and the Commission of the European Communities also attended as observers.

The subject of the Conference was extremely important. With the adoption in 1972 of the Oslo and London Conventions, some progress had been made in controlling pollution of the sea by dumping from ships. Various conventions established within the framework of the International Maritime Consultative Organization (IMCO) had as their object the control of other forms of sea pollution, especially that resulting from the discharge of oil or oily waste, whether accidental or deliberate. These IMCO conventions were in the process of being reviewed and strengthened (an international conference held in London resulted in the 1973 Convention for the Prevention of Pollution from Ships). But an important aspect of sea pollution was not covered by existing arrangements, viz. pollution which reaches the sea through rivers and estuaries, through pipelines or by direct discharge from the coast.

The Paris Convention, in so far as it might lead to a convention being concluded in this matter, would help to fill a major gap which existed at the time in the general system for marine protection.

The principal elements of the draft Convention which were discussed in Paris in September 1973 concerned:

(a) an undertaking by contracting parties to reduce or eliminate sea pollution caused by the discharge of substances listed in one or several annexes;

(b) the establishment, by the contracting parties, of a common monitoring network in the maritime zone covered by the Convention, namely the North Sea and the North-East Atlantic;

(c) the establishment of a commission (composed of representatives of the Contracting Parties), which might *inter alia* have the task of proposing the programmes of pollution-abatement in question, coordinating and exploiting the monitoring network; reviewing the list of substances in the annex or annexes; establishing environmental quality objectives to guide the general pollution-abatement programmes or sub-regional agreements which might be concluded between two or more of the contracting parties.

As noted above, the Commission of the European Communities participated as an observer at the first session of the Paris Conference.

On 12 November 1973 the Commission submitted to the Council a proposal for a Council decision concerning the participation of the European Economic Community in the negotiations for the conclusion of a Convention for the Prevention of Sea Pollution from Land-based Sources. In the light of this proposal, the Council authorized the Commission to participate jointly with the Member States concerned in the negotiations with the object of concluding the Convention, invited the Member States to work together with the Commission to achieve this objective, and reminded the Member States that they should adopt a common position on the essential provisions of the Convention.

The Diplomatic Conference on the Convention for the Prevention of Marine Pollution from Land-based Sources ended in Paris on 22 February 1974. On 3 March 1975 the Council adopted a decision concluding the Convention for the Prevention of Marine Pollution from Land-based Sources. The President of the Council was authorized to designate the persons in power to sign the Convention and to confer on them the powers they required to bind the Community.

The Community was to be represented by the Commission in the Commission established under the Convention. The Commission was to put forward the position of the Community in accordance with such directives as the Council might give it.

The Council also authorized the Commission to represent the Community in the working group entitled 'Interim Commission' established on the basis of resolution number III Annexed to the Final Act of the Convention.

In a Resolution on the same subject of 3 March 1975, the Council invited the Member States affected by the Convention for the Prevention of Marine pollution from Land-based Sources to sign the Convention as soon as possible, and in any case before 31 May 1975.

The Council considered it advisable to ensure coherent implementation of the undertakings entered into under this convention, of those which may result from the European convention for the protection of international watercourses against pollution, being drawn up by the Council of Europe, and, more generally, of the other commitments arising from the carrying out of the programme of action of the Communities on the environment.

On 23 June 1975, the Convention was signed on behalf of the Community. It entered into force in March 1978, all Community countries except Italy being contracting parties.

2. EUROPEAN CONVENTION FOR THE PROTECTION OF INTERNATIONAL WATERCOURSES AGAINST POLLUTION (STRASBOURG CONVENTION)

On 12 May 1969, the Consultative Assembly of the Council of Europe adopted Recommendation 555 under which it advised the Committee of Ministers to assign to a group of government experts the task of drawing up as quickly as possible a European Convention for the Protection of Fresh Water Against Pollution (based on a draft prepared by this Assembly).

After a series of meetings held between 1970 and 1973, an *ad hoc* Committee completed the draft text of the Convention.

Under the terms of this draft text the Member States of the Council of Europe would undertake in particular to:

(a) take all appropriate measures to reduce existing pollution and prevent new forms of water pollution;

(b) apply to international waterways either special standards or minimum standards. Some of the special standards were defined in an Annex to the Convention while others had yet to be worked out by a group of technical experts set up by a resolution of the Committee of Ministers. Such standards fixed the quality thresholds for the

water of international waterways according to the 'functions' assigned to the latter. These 'functions', which themselves corresponded to the various uses to which the waterways are put, would
continue to be laid down either by international commissions responsible for ensuring cooperation between contracting parties
whose territories are separated or crossed by the same international
waterway, or within the framework of intergovernmental agreements. The minimum standards were defined in an Annex to the
Convention.

The contracting parties, however, might request derogations (to
be defined in an Annex to the Convention) in respect of certain
waterways and certain parameters. These derogations would be determined by a second group of experts set up by a resolution of the
Committee of Ministers;

(c) enter into negotiations, if one of them so requested, for the conclusion of 'cooperation agreements' for the amendment of existing
agreements. If a contracting party did not enter into negotiations
within a reasonable time, any of the contracting parties concerned
would accordingly notify the Committee of Ministers of the Council
of Europe, which would place itself at the disposal of the contracting
parties for the purpose of seeking a procedure for arriving at a
satisfactory solution.

After an analysis of this text, the Commission took the view that there
was some measure of agreement between the Council of Europe's draft
Convention and the Community's programme as regards the objectives set
and the nature of the measures contemplated. Nevertheless, its general
economy and institutional aspects, plus the very fact that the object of the
Convention was covered by the Community's programme of action on the
environment, involved a risk that if the Convention were to remain as it
stood, the action taken at Community level might be affected.

Consequently, it was the Commission's view that certain amendments
should be made to the text in order to improve certain clauses and enable
the Community to sign this Convention.

The Commission accordingly proposed that the Council should call upon
Member States to reserve their position concerning the Convention until
the Commission had approached the Council of Europe with the view to
the amendment of the draft Convention in such a way as to enable the
European Economic Community as such as accede to this Convention side
by side with its Member States.

On 9 December 1974, the Commission reported to the Council that the
Secretary General of the Council of Europe had shown quite considerable
understanding with regard to the Community's position. This had facilitated
the achievement of an agreement *ad referendum* concerning the amendments
to be made in the draft convention.

In the light of the results of these negotiations and of the Council's decision on 7 November 1974 that the Community should participate in the Convention, the Commission proposed that the Council formally approve such participation.

At the present time, the Council has not yet been able to take this final decision since the European Convention for the Protection of International Watercourses against Pollution has not been finally adopted by the Ministers of the Council of Europe.

3. PROTECTION OF THE MEDITERRANEAN SEA AGAINST POLLUTION

In the context of the United Nations Environment Programme (UNEP), the countries bordering the Mediterranean were invited to attend an Intergovernmental Meeting, held in Barcelona from 28 January to 4 February 1975, on the protection of the Mediterranean.

During this meeting, a programme of action was adopted for the protection of the Mediterranean, involving:

(a) integrated planning for the development and management of the natural resources of the Mediterranean basin;

(b) a coordinated programme of research and continuous supervision, involving the exchange of information and appraisal of pollution levels and protection standards;

(c) a framework agreement on the protection of the Mediterranean marine environment, together with related protocols and technical annexes covering each of the principal sources of sea pollution;

(d) a study of the institutional and financial implications of this project.

On 7–11 April 1975 a meeting of legal experts was held in Geneva to examine the draft outline agreement and also the two draft Protocols on the prevention of dumping at sea by ships and aircraft and the control of pollution from hydrocarbons and other noxious substances as a result of accidents at sea. At the time of that meeting it was expected that one or more countries bordering the Mediterranean would take steps to propose draft Protocols on other sources of pollution, with special reference to the prevention of pollution from land-based sources.

By a decision of 8 December 1975, the Council authorized the Commission to take part, in the fields within the Community's competence, in the intergovernmental negotiations in Barcelona, with a view to enabling the Community to sign, if it so desired, an outline Convention, accompanied by protocols, relating to the preventing of marine pollution in the Mediterranean Sea.

In February 1976, a conference was held in Barcelona attended by the plenipotentiaries of the following States in the Mediterranean regions:

Cyprus, Egypt, France, Greece, Israel, Italy, Lebanon, Libya, Malta, Monaco, Morocco, Spain, Tunisia, Turkey and Yugoslavia, together with observers from the European Economic Community, the United Kingdom, the USA, the USSR and various international organizations.

The following texts were adopted by the Conference:

- the Convention on the protection of the Mediterranean Sea against pollution;
- a protocol on cooperation in the campaign against the pollution of the Mediterranean Sea by hydrocarbons and other harmful substances in the event of a critical situation arising;
- a protocol on the prevention of the pollution of the Mediterranean Sea by dumping from ships and aircraft;
- ten resolutions concerning the creation, objectives and tasks of a regional centre in Malta.

On 30 April the Commission submitted a communication to the Council in which it recommended that the Council should:

(a) approve the results of the negotiations which the representatives of the Community had conducted at Barcelona;
(b) decide to substantively approve the Convention on the protection of the Mediterranean Sea against pollution and of protocol on the prevention of the pollution of the Mediterranean Sea by dumping from ships and aircraft forthwith, and
(c) authorize the President of the Council to nominate the person or persons entitled to sign these agreements, subject to their conclusion, since the Convention must, under Article 24, be signed by the Community before 17 February 1977;
(d) adopt the proposal for a decision concerning the conclusion of these agreements by the Community.

On 13 September 1976 the signature, by the Community, took place in Madrid of the Convention on the Protection of the Mediterranean Sea against Pollution and of the protocol of the Convention of the Pollution of the Mediterranean Sea by dumping from Ships and Aircraft. The formal decision for concluding the Convention by the Community was taken by the Council on 15 July 1977.

On 14 June 1976, the Council authorized the Commission to participate in the negotiations concerning a protocol dealing with land-based sources of pollution of the Mediterranean Sea. On 18 December 1981 the Commission submitted a proposal for a decision concerning the conclusion of this protocol.

On 27 April 1978, the Commission proposed that the Council approve the accession of the EEC to a protocol concerning cooperation in combating pollution by oil and other harmful substances in cases of emergency. The

Council approved this proposal at its meeting of 19 December 1978. The Council subsequently adopted on 19 May 1981 a decision concerning the conclusion of this protocol which approved the protocol and authorized the President of the Council to deposit the instruments of accession.

On 31 March 1982 the Council authorized conclusion of the fourth protocol to the Barcelona Convention concerning special protected areas in the Mediterranean.

4. PROTECTION OF THE BALTIC SEA AGAINST POLLUTION

On 22 March 1974, in Helsinki, the Baltic states, namely: Denmark, Finland, the German Democratic Republic, the Federal Republic of Germany, Poland, Sweden and the USSR adopted a Convention on the Protection of the Marine Environment of the Baltic Sea Area. This Convention included amongst other things provisions for the control of pollution from the dumping of wastes.

On 21 June 1977, the Council adopted a decision authorizing the Commission to open negotiations with the countries bordering the Baltic Sea with a view to permitting the accession of the Community to the Baltic Convention.

5. PROTECTION OF THE WATERS OF THE RHINE BASIN AGAINST POLLUTION

The Action Programme on the Environment stressed the fact that the growing pollution of the waters of the Rhine and its tributaries was giving increasing cause for concern to people who used them or lived in the area. The Action Programme recalled various resolutions of the European Parliament on this subject as well as earlier communications from the Commission to the Council on the environment which also referred to the problem of the Rhine.

The Action Programme recalled that the signatory States of the Berne Convention, which set up an International Commission for the Protection of the Rhine against Pollution, took part in a ministerial conference held in The Hague on 25 and 26 October 1972 on the initiative of the Dutch Government. The Commission was represented at this conference as an observer.

Among a number of major decisions taken by this Conference was one on chemical pollution. The Conference decided that the International Commission would have the task of drawing up lists of materials in respect of which discharge must be prohibited, limited or made subject to certain conditions, of carrying out a survey to ascertain their source, and of working out a programme of action in stages which would be submitted to the governments for approval after one year.

At the Ministerial Conference on the protection of the Rhine against pollution, which was held in Bonn on 4 and 5 December 1973, the signatory States of the Berne Convention approved the general principles and in particular the three lists submitted by the International Commission covering substances whose discharges is to be prohibited, restricted or made subject to certain conditions. The Ministers agreed that, having regard to the Council of Europe declaration on the black list, their States would as far as possible prevent the discharge of substances on the black list into the waters of the Rhine catchment area. The Ministers instructed the International Commission to draft the text of an international agreement on the basis of those principles, if possible before the next Ministerial Conference, and to submit it to the States for approval.

This Conference, which the Commission of the European Communities attended as an observer, also agreed that in taking these measures, account would be taken of the work carried out by the European Communities, in accordance with their programme of action on the protection of the environment.

Several meetings were held by specialized working parties in the International Commission in order to prepare a draft convention on the protection of the Rhine against chemical pollution. This was discussed at a plenary meeting of the International Commission held in Amsterdam on 6 and 7 March 1975.

The draft convention discussed at the Amsterdam meeting made provision for the fixing of limit values for pollutants on a black list and quality objectives for pollutants on a grey and beige list. This draft also provided for the establishment of national inventories of discharges containing the substances appearing on the various lists.

In order to ensure that the limit values and quality objectives are adhered to, the draft Convention stipulated that the signatory States should make use of a variety of means according to the case in hand, viz., application of better techniques, implementation of laws and establishment of detailed programmes.

It was clear to the Commission that the draft convention currently being discussed by the International Commission for the Rhine made allowance, in addition to the provisions applicable to the individual Member States concerned, for the implementation of a number of measures already adopted or in the course of being adopted by the European Communities under its environment programme which gave the Communities authority in this field.

For this reason and also to ensure that action taken to prevent water pollution throughout the whole Community was consistent, the Commission felt that the European Economic Community as such should participate alongside the Member States concerned in the preparation of a Convention for the prospection of the Rhine against chemical pollution with a view to the Community's signing the Convention.

On 10 June 1975 the Commission submitted to the Council a draft Council decision to this effect. On 19 January 1976 the Commission received from the Council a mandate to negotiate.

The third ministerial Conference took place in Paris in April 1976. The Conference approved the draft Convention on the protection of the Rhine against chemical pollution and the participants indicated their intention to sign the Convention before the summer of 1976. The Ministers of the five riparian States also gave their agreement that the Community should become a contracting party. The Convention was therefore open for the signature of the Community.

On 17 September 1976 the Commission transmitted to the Council a draft decision relating to the conclusion by the Community of the Convention of the Protection of the Rhine against Chemical Pollution, and an additional Agreement to the Agreement signed in Berne on 29 April 1963 relating to the participation of the Community in the International Commission for the Protection of the Rhine against Pollution. This draft decision was approved by the Council on 25 July 1977. The process of ratification by contracting parties has been completed.

Action is now being taken to implement the Convention. A decision completing Annex IV of the Convention was adopted on 24 June 1982. The work goes in step with work on the dangerous substances directive of 4 May 1976.

The Ministers who met in Paris in April 1976 also pursued the discussions they had begun in The Hague in 1972 concerning the reduction of salt pollution. They decided to continue the negotiations, after making a certain amount of progress, and therefore met in Berne on 25 May 1976 with a view to reaching an agreement which would permit them to draw up a Convention for this kind of pollution.

The result of these ministerial deliberations was the approval, in its broad lines, of a draft agreement on concrete measures for a progressive reduction of the discharge of salt.

The Convention was signed on 3 December 1976. Because of opposition in Alsace to the scheme for injection of salt underground, the French Government was unable to ratify it. A revised scheme was agreed at a meeting of Ministers on 17 November 1981. It would not be operational before 1985. The Community is not a signatory of the Convention.

6. LAW OF THE SEA CONFERENCE

In 1970, the General Assembly of the United Nations decided that it would convene the Third United Nations Conference on the Law of the Sea. It made the Sea-Bed Committee the preparatory body for the Conference.

The subjects for the Conference were set out in General Assembly Resolution 2750 X (XXV), as follows:

- the establishment of an equitable international regime—including international machinery—for the area and the resources of the sea-bed and the ocean floor, and the subsoil thereof, beyond the limits of national jurisdiction;
- a precise definition of the area;
- a broad range of related issues including those concerning:
  - the regimes of the high seas, the continental shelf, the territorial sea (including the question of its breadth and the question of international straits) and contiguous zone;
  - fishing and conservation of the living resources of the high seas (including the question of the preferential rights of coastal States);
  - the preservation of the marine environment (including the prevention of pollution);
  - scientific research.

In 1973 the Assembly fixed as the goal of the Conference 'to adopt a convention dealing with all matters relating to the law of the sea'. It asked the Conference to bear in mind 'that the problems of ocean space are closely interrelated and need to be considered as a whole'.

The work of the Conference was carried on in three main Committees, each composed of all States taking part in the Conference.

The first Committee is concerned with a legal regime (body of rules) and machinery (world-wide authority) for the area of the sea-bed beyond the jurisdiction of individual States.

The Second Committee dealt with a wide range of issues concerning the law of the sea, from the territorial sea and the proposed economic zone to the rights of land-locked countries and the special problems of archipelagoes.

The third Committee is concerned with the preservation of the marine environment and with marine scientific research.

The texts which emerged from the negotiations in Committee III on marine pollution expand the obligation of the world community to protect endangered species and fragile ecosystems from pollution. They widen jurisdiction to establish ship-routing systems needed to protect the environment, clarify the right of a coastal state to obtain prompt notice of events that may result in pollution off its coast in adequate time to act, and remove certain restraints on the powers of the coastal state to enforce antipollution measures in the territorial sea and in the exclusive economic zone. They also clarify the right of the coastal state to establish and enforce discharge standards stricter than international standards for ships in innocent passage in the territorial sea.

The Community was invited to the Conference as an observer and was represented by a Commission delegation composed of officials from the principal departments concerned (the Legal Service, the Directorate-General for External Relations, the Directorate-General for Agriculture and the Environment Service).

On 2 June 1976, the Commission submitted a communication to the Council on the Third United Nations Conference on the Law of the Sea. In this communication the Commission proposed, *inter alia*:

● Adoption of a common position with regard to the provisions on pollution from vessels;
● adoption of common positions in order to ensure the coherent implementation of commitments to be entered into in the future Convention and of those undertaken by the Member States in the framework of the execution of the Community's environment programme.

The Commission also proposed that the Council authorize it to enter into negotiations at the Third United Nations Conference on the Law of the Sea with a view to having inserted in the International Convention of the Law of the Sea currently being drawn up by that Conference a clause enabling the European Economic Community to become a contracting party to the said Convention.

The draft Convention was finally adopted at the Eleventh Session on 30 April 1982. However, six Member States of the European Community abstained, while the United States voted against. Participation by the European Community in the Convention will require that a majority of Member States adhere to it.

7. PROTECTION OF THE NORTH SEA

(a) Prevention of pollution of North East Atlantic and North Sea from dumping

The Commission submitted on 19 December 1978 a proposal for a decision authorizing it to enter into negotiations leading to the eventual accession of the Community to the Oslo Convention on the Prevention of Pollution of the North East Atlantic and North Sea by dumping of wastes from ships and aircraft. Difficulties over the interpretation of the responsibilities of the member states and the Commission arose and the draft decision has not yet been adopted. The Commission sits as an observer in meetings of the Oslo Commission.

(b) Agreement on cooperation in dealing with oil pollution in the North Sea

The Bonn Agreement of 9 November 1969 provided for cooperation and exchange of information between signatories on dealing with oil pollution in the North Sea. At a meeting on 27–29 April 1982 it was agreed that the Community should accede.

8. MARITIME CONVENTIONS

The Commission in its Communication of 27 April 1978 on prevention and control of pollution of the sea especially by hydrocarbons asked the Council to agree to accelerate ratification of the following conventions.

(a) The MARPOL Convention

This was drawn up in 1973 following a conference in London on the prevention of marine pollution by ships organized by the Intergovernmental Maritime Consultative Organization (IMCO) and lays down tougher construction standards for tankers and designates special zones such as the Baltic and Mediterranean in which discharges of tank washings are banned. The Convention was amended at an IMCO conference in February 1978 with the drawing up of a protocol embodying additional safety rules requiring segregated ballast tanks in new tankers of over 20,000 tonnes and the use of onboard tank-washing systems for existing tankers of over 40,000 tonnes. It is expected to come into force shortly.

(b) The SOLAS Convention

The 1974 Convention sets out rules on construction, stability, radio communications, lifesaving equipment, etc., and like MARPOL was tightened up at the February 1978 IMCO conference by the addition of a protocol covering inspection of instruments and gear and the operation of ships in general. Some of the new provisions are intended to apply particularly to tankers including improved emergency steering requirements. These involve, among other things, mandatory steering failure alarms and duplication of elements in the steering system (except duplication of rudders themselves or the rudder drive).

(c) ILO Convention No. 147

The 1976 ILO Convention No. 147 (merchant shipping), is an outline convention under which signatories undertake to bring in laws requiring minimum social provisions and safety standards equivalent to rules already embodied in other ILO conventions covering officers, prevention of accidents, etc.

The French Government in 1980 took a major initiative in convening a meeting in Paris of the Ministers responsible for maritime safety in 13 countries of Western Europe for the purpose of devising concrete proposals for improving the safety of shipping and the prevention of pollution. As a result of that first meeting, in which the EEC Commission participated, proposals were drawn up for a harmonized and coordinated system for the inspection of foreign ships calling at European ports, for the purpose of detecting those which failed to meet the standards laid down in international conventions, securing the rectification of deficiencies and discouraging the operation of such vessels. These proposals were adopted at the second ministerial conference held in Paris on 26 January 1982, in the form of a Memorandum of Understanding on Port State Control, which came into effect on 1 July 1982. The countries which signed the memorandum include the nine Community maritime states plus Spain, Portugal, Norway, Sweden and Finland. The Commission played an active part in the conference and will be represented on the committee to be set up to monitor the operation of the scheme.

9. LONG RANGE TRANSBOUNDARY AIR POLLUTION

One outcome of the Helsinki Conference on Security and Cooperation in Europe was to call for cooperation to control air pollution and its effects including long range transport of air pollutants. Negotiations therefore started in the United Nations Economic Commission for Europe at Geneva on a Convention on Long Range Transboundary Air Pollution. The Community participated in these negotiations and signed along with other Member States the Convention on 14 November 1979. The Council subsequently by their decision of 11 June 1981 approved the conclusion of the Convention. The Convention provides for an undertaking by the Contracting Parties to develop the best policies and strategies to control air pollution, to cooperate in research and development and to exchange information. In the proposals for the 3rd Action Programme the Commission indicated that a policy would have to be devised which initially would stabilize and thereafter gradually reduce total emissions by establishing emission standards, where necessary, for certain sources. This would apply notably to large, fixed sources with high stacks which distribute pollutants over a wide radius. This action would form part of the Commission's contribution to the effort to resolve the acid-rain and transboundary pollution problems being organized under this Convention.

10. COOPERATION WITH THIRD COUNTRIES

The Commission has exchanged letters regarding cooperation in environmental matters with the United States, Canada, Japan, Norway, Sweden, Switzerland and Austria.

The purpose of such exchanges of information is to enable both the Community and the non-Member States to keep each other up to date on the progress of their respective work, thus avoiding unnecessary duplication and, by acting in harmony, forestalling any problems which might arise from differences in the assessment and handling of environmental issues.

Annex I
Environment Information
Agreement

ANNEX I (a)
Council

Agreement of the representatives of the governments of the Member States* meeting in Council of 5 March 1973 on information for the Commission and for the Member States with a view to possible harmonization throughout the Communities of urgent measures concerning the protection of the environment.

THE REPRESENTATIVES OF THE GOVERNMENTS OF THE MEMBER STATES, MEETING IN COUNCIL

Whereas the reduction of pollution and nuisances is of particular importance for the European Communities; whereas measures must be taken as a matter of urgency to combat such pollution and nuisances;

Whereas the Commission has proposed to the Council an action programme in this respect which must aim to maintain and, wherever possible, to improve the quality of the environment in the territory of the Member States of the Community;

Whereas measures in this sector are being prepared in most of the Member States;

Whereas some of these measures, if they are not harmonized, could affect the functioning of the Common Market and the implementation of the Communities programme for the reduction of pollution and nuisances and the protection of the natural environment;

Whereas, nevertheless, the pursuit of harmonization must not delay the adoption of essential measures for better protection of the environment;

Whereas knowledge of the intentions of Member States in this respect must be available to the Commission and the Member States, especially to enable the latter to propose Community measures where appropriate;

Whereas it is therefore necessary to establish a procedure for giving information concerning the intentions of the Governments of the Member States and concerning measures which they have in draft; especially where such measures are likely to

* The Representatives of the Governments of the Member States, meeting in Council, take note that this Agreement is a gentlemen's agreement.

affect the functioning of the common market and the implementation of the Communities' programme for the reduction of pollution and nuisances and the protection of the natural environment;

Whereas such information must be given as early as possible before the entry into force of the measures envisaged;

Whereas the Governments of the Member States must however be able, by way of exception, to take immediate action at national level when this is urgently necessary for reasons of safety or health;

HAVE AGREED AS FOLLOWS:

In order to ensure that the Commission and the Governments of the Member States are kept informed and in order to allow the Commission, where appropriate, to submit suitable proposals to the Council;

1. The Commission shall be informed as soon as possible of any draft legislative, regulatory or administrative measures and of any international initiative concerning the protection or improvement of the environment which

   ● may directly affect the functioning of the common market, or

   ● are relevant to the Communities' programme for the reduction of pollution and nuisances and the protection of the natural environment,

   or

   ● are of particular interest to the Communities and the Member States from the point of view of the protection of public health or of the natural environment, particularly where these measures may have repercussions for other Member States.

   The Government of the Member States take note that the Commission will, as soon as possible, communicate to the Governments of the Member States all information acquired pursuant to this Agreement.

2. The legislative, regulatory or administrative measures referred to in item 1, which are liable to have a direct effect on the functioning of the common market shall only be adopted if the Commission does not notify the Governments concerned, within two months of receiving such information, of its intention to submit to the Council proposals to adopt Community measures on this subject. Such proposals must take into account the aims of the national measures in question from the point of view of environmental protection.

   However, if the Commission does not submit to the Council a proposal within five months of receipt of such said information, the Government concerned may proceed immediately with the proposed measures. The same shall apply if the Council has not acted on the proposal from the Commission within five months of its receipt.

3. In appropriate cases (2), the procedure described in item 2 will be extended to draft measures liable to affect the implementation of the Communities' programme, as adopted by the Council, for the reduction of pollution and nuisances and to the protection of the natural environment.

4. Notwithstanding the foregoing and by way of exception, legislative, regulatory or administrative measures may be adopted if these are urgently necessary for serious reasons of safety or health. The Governments of the Member States will

immediately communicate the texts concerning such measures to the Commission which will transmit them to the Governments of the other Member States as soon as possible.

5. The Governments of the Member States will coordinate their views on any international initiative in respect of the environment likely to affect the functioning of the common market or the implementation of those parts of the Communities' programme for the reduction of pollution and nuisances and the protection of the natural environment to which the procedure laid down in item 2 applies by virtue of item 3, without prejudice to the application of the Treaties and in particular of Article 113 and 116 of the Treaty establishing the European Economic Community.

ANNEX I (b)

Agreement of the representatives of the governments of the Member States of the European Communities, meeting in Council of 15 July 1974 supplementing the agreement of 5 March 1973 on information for the Commission and for the Member States with a view to possible harmonization throughout the Communities of urgent measures concerning the protection of the environment.

THE REPRESENTATIVES OF THE GOVERNMENTS OF THE MEMBER STATES OF THE EUROPEAN COMMUNITIES MEETING IN COUNCIL

Whereas an Agreement on information for the Commission and for the Member States with a view to possible harmonization throughout the Communities of urgent measures concerning the protection of the environment, was concluded on 5 March 1973; whereas the application of point 3 of this Agreement should be more precisely defined, as provided for in the footnote relating to this point,

HAVE AGREED AS FOLLOWS:

Initially, the appropriate cases referred to in point 3 of the aforesaid Agreement shall concern the draft legislative, regulatory or administrative measures, i.e. measures of a binding nature, which are liable to affect the implementation of the programme of action of the European Communities on the environment, approved on 22 November 1973, wherever it is laid down that, for the implementation of this programme, the Commission shall submit proposals for the relevant measures of the Communities inasmuch as the latter provisions are required to take the form of Regulations or Directives.

Annex II
Environmental Protection
Terminology

Preliminary note:

Some of the terms defined below may have application outside the field of environmental protection. The definitions set out here are not concerned with these.

1. CRITERIA

1.1 The term 'criterion' signifies the relationship between the exposure of a target to pollution or nuisance, and the risk and/or the magnitude of the adverse or undesirable effect resulting from the exposure in given circumstances.

1.2 'Target' means man or any component of the environment actually or potentially exposed to pollution or nuisance.

1.3 The 'exposure' of a target, envisaged in this relationship, should be expressed as numerical values of concentration, intensity, duration or frequency.

1.4 'Risk' is the probability of occurrence of adverse or undesirable effects arising from a given exposure to one or more pollutants or nuisances considered alone or in combination with others.

1.5 The 'adverse or undesirable effect' envisaged in this relationship may be a direct or indirect, immediate or delayed, simple or combined action on the target. The risk and the magnitude of this effect should be expressed, whenever possible, in quantitative terms.

1.6 The methods of evaluating the parameters describing exposure and adverse or undesirable effects should be harmonized to ensure comparability of the results from studies and research on criteria.

2. QUALITY OBJECTIVES

2.1 The 'quality objective' of an environment refers to the set of requirements which must be fulfilled at a given time, now or in the future, by a given environment or particular part thereof.

2.2 In setting this objective, the following are taken into account:

(a) a 'basic protection level' such that man or another target is not exposed to any unacceptable risk.

(b) a 'no-effect level' such that no identifiable effect will be caused to the target.

These two levels are determined on the basis of the criteria described above. Due allowance is also made for the specific regional conditions, the possible effects on neighbouring regions, and the intended use.

3. ENVIRONMENTAL PROTECTION STANDARDS

3.1. 'Standards' are established in order to limit or prevent the exposure of targets and can thus be a means of achieving or approaching quality objectives. The standards are directly or indirectly addressed to the responsible individuals or bodies and set levels for pollution or nuisance that must not be exceeded in an environment, target, product, etc. They may be established by means of laws, regulations or administrative procedures or by mutual agreement or voluntary acceptance.

3.2 Standards include:

3.2.1 'Environmental quality standards' which, with legally binding force, prescribe the levels of pollution or nuisance not to be exceeded in a given environment or part thereof.

3.2.2 'Product standards' (the term product is used here in its broadest meaning) which

- set levels for pollutants or nuisance which are not to be exceeded in the composition or the emission of a product:
- or specify properties or characteristics of design of a product;
- or are concerned with the way in which products are used.*

Where appropriate, product standards include specifications for testing, packaging, marking and labelling products.

3.2.3 Standards for fixed installations, sometimes called 'process standards' such as:

(a) 'emission standards', which set levels for pollutants or nuisances not to be exceeded in emissions from fixed installations;

(b) 'installation design standards', which determine the requirements to be met in the design and construction of fixed installations in order to protect the environment,

(c) 'operating standards', which determine the requirements* to be met in the operation of fixed installations in order to protect the environment.

3.3 In some cases, it may be advisable to set standards even when it has not yet been possible to formulate the relevant criteria and quality objectives.

---

* Such methods of use and specifications may be issued in the form of 'codes of practice'

4. GENERAL

In all instances, as knowledge develops, criteria, objectives and standards will need to be periodically reviewed and, where appropriate, altered.

Annex III
Adaptation to Technical Progress Procedure

COUNCIL RESOLUTION

of 15 July 1975

on the adaptation to technical progress of Directives or other Community
rules on the protection and improvement of the environment

THE COUNCIL OF THE EUROPEAN COMMUNITIES:

Having regard to the draft of the Commission;

Having regard to the Opinion of the European Parliament;

Having regard to the Opinion of the Economic and Social Committee;

Whereas in order to carry out the programme of action of the European Communities on the environment it will be necessary among other things to adopt Directives or other Community rules for approximating the laws, regulations and administrative provisions of the Member States, and in the absence of national legislation to establish Community rules;

Whereas it may sometimes be necessary to adapt certain measures contained in these Directives or other Community rules to scientific and technical progress, particularly in the environment field, in accordance with an *ad hoc* procedure;

A. Adopts to this end as a solution in principle:

- the establishment of committees composed of representatives of the Member States and chaired by a representative of the Commission;
- the insertion in the Directives or other Community rules of the following provisions;

'*Article.* . .

1. Where the procedure laid down in this Article is to be followed matters shall be referred by the Chairman, either on his own initiative or at the request of the representative of a Member State, to the Committee. . ., hereinafter called the 'Committee'.

2. The representative of the Commission shall submit to the Committee a draft of the measures to be adopted. The Committee shall deliver its Opinion on the draft within a time limit which may be determined by the Chairman according to the urgency of the matter. It shall decide by a

majority of 41 votes, the votes of the Member States being weighted as provided for in Article 148 (2) of the Treaty. The Chairman shall not vote.

3. (a) The Commission shall adopt the measures envisaged where these are in accordance with the Opinion of the Committee;

   (b) When the measures envisaged are not in accordance with the Opinion of the Committee, or if no opinion is adopted, the Commission shall, without delay, propose to the Council the measures to be adopted. The Council shall act by a qualified majority.

   (c) If, within three months of the proposal being submitted to it, the Council has not acted, the measures proposed shall be adopted by the Commission.';

B. Provides that in cases which in the opinion of the Commission are of special importance, the Commission shall submit proposals directly to the Council, and the Council shall act on these by a qualified majority;

C. Agrees that in each Directive or other Community rule it shall be specified for which provisions thereof the procedure set out above shall be invoked.

D. Agrees, in addition, that at the end of a period of 18 months from the initial application of the procedures set out above, it will investigate, at the request of a Member State, on a proposal from the Commission, and in the light of experience, whether there is a need to amend these procedures.

Annex IV
EEC Legislation and Proposals in the Field of Pollution Control

1. PROPOSALS ADOPTED BY THE COUNCIL OF MINISTERS

Title	Date of adoption *	Official Journal Reference
1. Agreement of the Representatives of the governments of the Member States on information for the Commission and for the Member States with a view of possible harmonization throughout the Community of urgent measures concerning protection of the environment	5.3.73	OJ C 9 (15.3.73)
2. Directive on detergents	22.11.73	OJ L 347 (17.12.73)
3. Directive on the method of control of the bio-degradability of anionic surfactants.	22.11.73	OJ L 347 (17.12.73)
4. Declaration of the Council of the European Communities and of the representatives of the Governments of the Member States meeting in the Council of 22 November 1973 on the programme of action of the European Communities on the environment.	22.11.73	OJ C 112 (20.12.73)
5. Directive relating to certain parts and characteristics of wheeled agricultural or	4.3.74	OJ L 84 (28.3.74)

* The date of adoption given is the date of *formal adoption* and not necessarily the date on which the Council of Ministers agreed the proposal. Where a proposal has been agreed by the Council but not yet formally adopted, the date on which agreement was reached is given in parentheses.

forestry tractors (permissible limits for
overload)

6. Adaptation to technical progress of the Council directive of 20.3.70 concerning measures to be taken against air pollution by gases from positive ignition engines of motor vehicles.	28.5.74	OJ L 159 (15.6.74)
7. Amendment to directive of 2.8.72 on measures to be taken against emissions of pollutants from diesel engines for use in motor vehicles		OJ L 215 (6.8.74)
8. Agreement of the Representatives of the Governments of the Member States supplementing the agreement of 5.3.73 on information for the Commission and for the Member States with a view to possible harmonization throughout the Communities of urgent measures concerning the protection of the environment.	15.7.74	OJ C 86 (20.7.74)
9. Resolution on energy and the environment	3.3.75	OJ C 168 (25.7.75)
10. Resolution on the Convention for the Prevention of Marine Pollution from Land-based Sources	3.3.75	OJ C 168 (25.7.75)
11. Decision concluding the Convention for the Prevention of Marine Pollution from Land-based Sources (the Paris Convention). Decision concerning Community participation in the Interim Commission established on the basis of resolution No 111 of the Convention for the Prevention of Marine Pollution from Land-based Sources.	3.3.75	OJ L 194 (25.7.75)
12. Recommendation to the Member States regarding cost allocation and action by public authorities on environmental matters (applying the 'polluter pays' principle).	3.3.75	OJ L 194 (25.7.75)
13. Directive concerning the quality required of surface water intended for the abstraction of drinking water in the Member States	16.6.75	OJ L 194 (25.7.75)
14. Directive on the disposal of waste oils.	16.6.75	OJ L 194 (25.7.75)
15. Resolution concerning a revised list of second category pollutants to be studied as part of the programme of action of the European Communities on the environment.	24.6.75	OJ C 168 (25.7.75)

16. Decision establishing a common procedure   24.6.75   OJ L 194 (25.7.75)
    for the exchange of information between
    the surveillance and monitoring networks
    based on data relating to atmospheric
    pollution caused by certain compounds
    and suspended particulates.

17. Decision adopting an indirect action   26.6.75   OJ L 178 (19.7.75)
    programme for the management and
    storage of radioactive waste.

18. Directive on waste (framework directive).   15.7.75   OJ L 194 (25.5.75)

19. Resolution on the adaptation to technical   15.7.75   OJ C 168 (25.7.75)
    progress of directives or other Community
    rules on the protection and improvement
    of the environment.

20. Directive on the approximation of the laws   24.11.75   OJ L 307 (27.11.75)
    of the Member States relating to the
    sulphur content of certain liquid fuels (gas
    oil).

21. Directive concerning the quality of bathing   8.12.75   OJ L 31 (5.2.76)
    waters.

22. Decision authorising the Commission to   8.12.75   Not published in OJ
    participate in the negotiation of an outline
    Convention for the prevention of marine
    pollution in the Mediterranean.

23. Decision establishing a common procedure   8.12.75   OJ L 31 (5.2.76)
    for the setting-up and constant up-dating
    of an inventory of sources of information
    on the environment

24. Decision on the participation of the EEC   20.1.76   Not published in OJ
    in the Convention for the Prevention of
    Chemical Pollution of the Rhine.

25. Decision adopting a research and training   15.3.76   OJ L 74 (20.3.76)
    programme (1976–80) for the European
    Atomic Energy Community in the field of
    biology and health protection ('radiation
    protection' programme).

26. Decision adopting a research programme   15.3.76   OJ L 74 (20.3.76)
    (1976–80) for the European Economic
    Community in the environmental field
    (indirect action).

27. Directive on the disposal of PCBs and   6.4.76   OJ L 108 (26.4.76)
    PCTs

28. Directive on pollution caused by certain   4.5.76   OJ L 129 (18.5.76)
    dangerous substances discharged into the
    aquatic environment of the Community.

29. Directive on the approximation of the laws   27.7.76   OJ L 262 (27.9.76)
    of the Member States restricting the

marketing and use of certain dangerous
substances and preparations. PCBs and
PCTs

30. Second amendment to the directive of   30.11.76   OJ L 32 (3.2.77)
    20.3.70 concerning measures to be taken
    against air pollution by gases from positive
    ignition engines of motor vehicles

31. Directive modifying the Council Directive   8.3.77   OJ L 66 (12.3.77)
    of 6.2.70 on the approximation of the laws
    of the Member States relative to the
    permissible sound level and the exhaust
    system of motor vehicles.

32. Directive on screening the population for   29.3.77   OJ L 105 (28.4.77)
    lead.

33. Resolution on the continuation and   17.5.77   OJ C 139 (13.6.77)
    implementation of a European Community
    policy and action programme on the
    environment.

34. Decision authorising the Commission to   21.6.77   Not published in OJ
    open negotiations for the accession of the
    EEC to the Helsinki Convention of
    22.3.74 on the protection of the marine
    environment of the Baltic Sea area.

35. Directive on the approximation of the laws   28.6.77   OJ L 220 (29.8.77)
    of the Member States relating to the
    measures to be taken against the emission
    of pollutants from diesel engines for use in
    wheeled agricultural or forestry tractors.

36. Decision concluding the Convention on   25.7.77   OJ L 240 (19.9.77)
    the Protection of the Rhine against
    Chemical Pollution.

37. Decision concluding the Convention on   25.7.77   OJ L 240 (19.9.77)
    the Protection of the Mediterranean and
    the Protocol on prevention of pollution
    from shipping and aircraft dumping

38. Decision establishing a dumping procedure   12.12.77   OJ L 334 (24.12.77)
    for the exchange of common information
    on the quality of surface fresh water in the
    Community.

39. Directive on the reduction of pollution   20.2.78   OJ L 54 (25.2.78)
    arising from the production of titanium
    dioxide.

40. Directive on toxic and dangerous wastes.   20.3.78   OJ L 84 (31.3.78)

41. Recommendation on fluorocarbons in the   30.5.78   OJ C 133 (7.6.78)
    environment

42. Resolution on measures for the   26.6.78   OJ C 162 (8.7.78)
    prevention, control and reduction of

pollution caused by accidental discharges
of hydrocarbons into the sea.

43. Directive concerning the classification,   26.6.78   OJ L 206 (29.7.78)
    packaging and labelling of dangerous
    preparations (pesticides).

44. Directive on the protection of the health   29.6.78   OJ L 197 (22.7.78)
    of workers exposed to vinyl chloride
    monomer (VCM).

45. Directive relating to the lead content of   29.6.78   OJ L 197 (22.7.78)
    petrol

46. Directive on the quality requirements for   18.7.78   OJ L 222 (14.8.78)
    waters capable of supporting fresh-water
    fish.

47. Directive on the permissible sound level   23.11.78   OJ L 349 (13.12.78)
    and exhaust system of motor cycles

48. Directive relating to the quality of water
    for human consumption

49. Recommendation to Member States   19.12.78   OJ L 5 (9.1.79)
    regarding methods of evaluating the cost
    of pollution control to industry

50. Directive on the approximation of the laws   19.12.78   OJ L 33 (8.2.79)
    of the Member States relating to the
    determination of the noise emission of
    construction plant and equipment.

51. Directive concerning the marketing and   21.12.78   OJ L 33 (8.2.79)
    use of plant protection products containing
    certain active substances

52. Directive amending for the sixth time   18.9.79   OJ L 259 (15.10.79)
    Directive 67/648/EEC of 27.6.67 on the
    approximation of laws, regulations and
    administrative provisions relating to the
    classification, packaging and labelling of
    dangerous substances

53. Directive concerning measurement   9.10.79   OJ L 271 (29.10.79)
    methods and frequency of sampling and
    analysis of surface waters intended for the
    abstraction of drinking water in Member
    States

54. Directive on the quality required for   30.10.79   OJ L 281 (10.11.79)
    shellfish waters

55. Directive concerning the limitation of   20.12.79   OJ L 18 (24.1.80)
    noise from sub-sonic aircraft

56. Directive on the protection of   17.12.79   OJ L 20 (26.1.80)
    groundwater against pollution caused by
    certain dangerous substances

57. Decision concerning chlorofluorocarbons   26.3.80   OJ L 90 (3.4.80)
    in the environment

58. Directive on the quality of water for human consumption	15.7.80	OJ L 229 (30.8.80)
59. Directive on air quality limit values and guide values for sulphur dioxides and suspended particles.	15.7.80	OJ L 229 (30.8.80)
60. Resolution concerning transfrontier air pollution from sulphur dioxides and suspended particulates	15.7.80	OJ C 222 (30.8.80)
61. Decision adopting a sectoral research and development programme in the field of environment (environmental protection and climatology)—indirect and concentrated actions—(1981 to 1985)	3.3.81	OJ L 24 (11.4.81)
62. Decision concerning the conclusion of the protocol relating to cooperation in combatting pollution of the Mediterranean Sea by hydrocarbons and other harmful substances in cases of emergency.	19.5.81	OJ L 162 (19.6.81)
63. Decision concerning the conclusion of the Convention on long range transfrontier pollution	11.6.81	OJ L 171 (27.6.81)
64. Decision setting up a Community system of information for the control and reduction of pollution from oil spills at sea	3.12.81	OJ L 355 (10.12.81)
65. Recommendation concerning the re-use of waste paper and use of recycled paper	3.12.81	OJ L 355 (10.12.81)
66. Directive on limit volumes and quality objectives for measuring discharges from the chlor-alkali electrolysis industry.	22.3.82	OJ L 81 (27.3.82)
67. Directive on the hazards of major accidents from certain industrial activities.	24.6.82	OJ L 230 (5.8.82)
68. Decision concerning conclusion of protocol on special protection areas in the Mediterranean.	31.3.82	OJ L 106 (21.4.82)
69. Decision on approximation of laws of Member States relating to methods of testing biodegradability of non-ionic surfactants and amending directive 73/404/EEC of 22 November 1973.	31.3.82	OJ L 108 (22.4.82)
70. Decision amending directive 73/405/EEC concerning control of biodegradability of anionic surfactants	31.3.82	OJ L 108 (22.4.82)
71. Directive on a quality standard for lead in air	24.6.82	OJ L 378 (31.12.82)
72. Directive on methods for surveillance and monitoring of discharges of waste from the titanium dioxide industry	24.6.82	OJ L 378 (31.12.82)

73. Decision establishing a reciprocal exchange of information and data from networks and individual stations measuring air pollution within member States.	24.6.82	OJ L 210 (19.7.82)
74. Decision on the consolidation of precautionary measures on chlorofluorocarbons in the environment.	24.6.82	OJ L 329 (24.11.82)
75. Decision completing Annex IV of the Convention on the protection of the Rhine against chemical pollution	24.6.82	OJ L 210 (19.7.82)
76. Resolution on the continuation and implementation of a European community policy and action programme on the enviroment.	17.12.82	OJ C 56 (1.3.83)

2. PROPOSALS ADOPTED BY THE COMMISSION

Title	Date of adoption	Official Journal Reference
1. Recommendation to Member States invited to take part in the intergovernmental meeting in Barcelona.	19.12.75	OJ L 9 (16.1.76)
2. Decision concerning the setting up of a waste management committee.	21.4.76	OJ L 115 (1.5.76)
3. Objective evaluation of the risk to human health from pollution by some persistent organo-chlorine compounds (Communication to the Council).	29.7.76	OJ C 212 (9.9.76)
4. Determination of criteria for noise (Communication to Council).	3.12.76	COM (76) 646
5. Decision concerning the setting up of a Scientific Advisory Committee to examine the toxicity and ecotoxicity of chemical compounds.	28.6.78	OJ L 198 (22.7.78)
6. Decision concerning the setting up of an advisory committee on the control and reduction of pollution from oil spills at sea	25.6.80	OJ L 188 (27.7.80)
7. Decision defining the criteria governing the information for the inventory of chemical substances to be provided to the Commission by Member States.	11.5.81	OJ L 167 (24.6.81)

3. PROPOSALS SUBMITTED BY THE COMMISSION AND NOW UNDER DISCUSSION IN THE COUNCIL

Proposal	Date of proposal	Official Journal Reference
1. Decision concluding a European Convention for the Protection of International Watercourses Against Pollution.	11.12.74	OJ C 99 (2.5.75)
2. Directive on the approximation of the laws of the Member States concerning the classification, packaging and labelling of pesticides.	20.12.74	OJ C 40 (20.2.75)
3. Directive on the approximation of the laws of the Member States relating to the permissible sound level for pneumatic concrete-breakers and jackhammers.	31.12.74	OJ C 82 (14.4.75)
4. Directive on the reduction of water pollution caused by wood pulp mills in the Member States.	20.1.75	OJ C 99 (2.5.75)
5. Directives on permissible sound levels for current generators for welding, tower cranes and current generators for power supply.	30.12.75	OJ C 54 (8.3.76)
6. Directive concerning the dumping of wastes at sea	12.1.76	OJ C 40 (20.2.76)
7. Directive concerning the marketing of EEC-accepted plant protection products.	4.8.76	OJ C 212 (9.9.76)
8. Directive on noise levels for air compressors	5.4.78	OJ C 94 (19.4.78)
9. Directive on noise levels for lawn mowers	18.12.78	OJ C 86 (2.4.79)
10. Decision on negotiations for community accession to the Oslo Convention on dumping at sea.	19.12.78	
11. Directive on limit values for discharges to the environment of aldrin, dieldrin and endrin. Directive on quality objectives for aldrin, dieldrin and endrin.	16.5.79	OJ C 146 (12.6.79)
12. Directive on assessment of the environmental effects of certain public and private projects.	16.6.80	OJ C 169 (9.7.80)
13. Directive on noise levels for dozers, loaders and excavators.	28.11.80	OJ C 356 (31.12.80)
14. Directive on fixing limit values for discharge of cadmium to the aquatic	17.2.81	OJ C 118 (21.5.81)

environment and quality objectives for the
level of cadmium.

15. Directive on containers of liquids for   23.4.81   OJ C 204 (13.8.81)
    human consumption.

16. Directive amending Council directive 80/   28.9.81   OJ C 276 (28.10.81)
    51/EEC of 20.12.79 on noise levels for sub
    sonic aircraft.

17. Directive on noise levels for helicopters.   13.10.81   OJ C 275 (27.10.81)

18. Decision concluding the protocol on the   18.12.81   OJ C 4 (8.1.82)
    Barcelona Convention on protection of the
    Mediterranean sea from land based
    pollution.

19. Directive amending Council Directive 70/   5.4.82   OJ C 2
    220/EEC on approximation of laws of   (COM (82)170)
    Member States relating to measures to be
    taken against air pollution by gases from
    positive ignition engines of motor vehicles.

Annex V
Selected Bibliography*

1. GENERAL

General Report on the Activities of the European Communities
5th General Report 1971 and subsequent Annual Reports (all contain a section on the environment

Terminologie de l'environnement/Terminology of the environment.
(European Parliament). *Beiträge zur Umweltgestaltung.* Heft B 9 1974.
Erich Schmidt Verlag. Edition a feuillets mobiles, pag. diff. (f.i.e.d.n.)†

A Community programme for the Environment
1972. 76pp.
Supplement to the Bulletin of the EC, no. 5/72.
(d.e.f.i.n.)

A Community programme for the Environment
1973. 70pp.
Supplement to the Bulletin of the EC, no. 3/73.
(d.e.f.i.n.)

A Community programme for the Environment
1976. 70pp.
Supplement to the Bulletin of the EC no 5/76.
(d.e.f.i.n.)

Recent progress in evaluating the effects of pollution on health
Act – International Symposium. Paris 24–28 June 1974.
(Organised jointly by Commission of the European Communities, United States Environmental Protection Agency and World Health Organisation).
1975 Vol. I   :   473pp.   + index. (mult.)
    Vol. II : pp.   474–1188 + index. (mult.)
    Vol. III : pp. 1189–1872 + index. (mult.)

* The reference EUR indicates that the item has been published by the Commission of the European Communities.
Reports listed here which have been published by other publishers were sponsored by the Commission.

† Signifies language in which item is available.

Vol. IV : pp. 1873–2522 + index. (mult.)
(mult. EUR 6851)

The Law and Practic Relating to Pollution Control in the Member States of the European Communities. 1983.
Published by Graham & Trotman Ltd, London, at the invitation of the Commission of the European Communities.
Series co-ordinated by Environmental Resources Limited.
Each hardback volume is available with a paperback updating supplement.

1. *The Law and Practice Relating to Pollution Control in the Member States of the European Communities – A Comparative Survey.* 388 pp.
2. *The Law and Practice Relating to Pollution Control in Belgium and Luxembourg.* 386pp.
3. *The Law and Practice Relating to Pollution Control in Denmark.* 234pp.
4. *The Law and Practice Relating to Pollution Control in France.* 176pp.
5. *The Law and Practice Relating to Pollution Control in the Federal Republic of Germany.* 336pp.
6. *The Law and Practice Relating to Pollution Control in Ireland.* 244pp.
7. *The Law and Practice Relating to Pollution Control in Italy.* 244pp.
8. *The Law and Practice Relating to Pollution Control in the Netherlands.* 150pp.
9. *The Law and Practice Relating to Pollution Control in the United Kingdom.* 420pp.
10. *The Law and Practice Relating to Pollution Control in Greece.* 178pp.

Levels of Pollution of the Environment by the Principal Pollutants.
J. Bouquiaux. 1977. 116pp.
(e. EUR 5730).

Results of environmental measurements in the Member States of the European Community for air-deposition-water-milk
1979. 258pp.
(da.d.e.f.i.n. EUR 7032)

Environmental impact of energy strategies within the EEC
1980. 169pp
(e. EUR 6571)

CEC harmonzation of methods for measurement of NO$_x$: Intercalibration of measuring equipment
1981. 82pp
(e. EUR 7865)

2. POLLUTION CAUSED BY METALS

Environmental health aspects of lead of the international symposium at Amsterdam (Netherlands) October 2–7, 1972.
1973.
(d.e.f. EUR 5004).
Van Wambeke, P.

Géochimie du mercure et pollution de l'environnement
1975. 35pp.
(f. EUR 5241).

The Behaviour of Chromium in aquatic and terrestrial food chains
1975. 81pp.
(e. EUR 5375)

Sabbioni, E., Girardim, F., Harafante, E.
A systematic study of biochemical effects of heavy metal pollution
Programme of the research and preliminary results on cadmium and lead.
1975. 30pp. + fig.
(e. EUD 5333).

*Problems presented by the contamination of man and the environment by mercury
and cadmium.*
Act – International symposium, 3/4/5 July 1973.
1974. 696pp.
(d.e.f.i.n. EUR 5075).

Le mercure, le cadmium et le chrome aux Pays-Bas
1973.
(f. EUR 5006).

*Etude sur les sources et les volumes de résidus et déchets solides, liquides et gazeux
de métaux lourds rejetés dans le milieu ambiant en RF d'Allemagne et en France*
1973.
(d.f. EUR 5005).

*Aspects of the industrial use of cadmium. No. 1. Production, consumption and uses
of cadmium in the European Community*
142pp
(e. EUR 6636)

*Aspects of the industrial use of cadmium. No. 2. Survey of industrial emission of
cadmium in the European Economic Community*
51pp
(e. EUR 6636)

*Aspects of the industrial use of cadmium. No. 3. Methods and costs of preventing
cadmium emissions*
29pp
(e. EUR 6636)

The Pathway of Mercury in Europe (A system dynamics model)
J. Randers, R. H. Van Enk, P. Zegers
1977. 48pp.
(e. EUR 5670)

*Future average mercury content of air, soil and river sediment in the EEC and in
the world's oceans.*
1979.
(e. EUR 6023)

*The Pathway of Mercury and the impact of different environmental policies on the
future mercury content in soil, air and sediments in Europe.*
R. H. Van Enk and P. Zegers.
1977. 62pp.

(e. EUR 5683).

Environment and site problems in the uranium ore mining and processing industry
20pp.
(e. EUR 6586)

3. POLLUTION CAUSED BY PESTICIDES

*The content of organo-halogen compounds detected between 1968 and 1972
in water, air and foodstuffs and the methods of analysis used in the nine
Member States of the European Community.*
Report of a working group of experts.
Rapporteur: R. Mestres – Université de Montpellier.
1974.
(e. ref. V/F/1630/74).

*Problems presented by the contamination of man and the environment by
pesticides and persistent organohalogen compounds.*
European Colloquium, Luxembourg, 14–16 May 1974.
1975. 647pp. + Glossary
(mult. EUR 5126)

*Pesticide residues in human fat and human milk in the nine Member States
of the European Community* (1969–1973).
Report of a working group of experts.
Rapporteur: G. L. Gatti.
1974.
(e. ref. V/F/1629/74).

4. POLLUTION OF THE AIR

Technical measures against atmospheric pollution in the steel industry: Report and information on research carried out with financial aid from ECSC.
1973. 80pp.
(d.e.f.i.n.)

Jost, D.
Geräte und Messtechniken für die Feststellung der Luftverschmutzung
1974. 104pp.
(d. EUR 5135)

Commission of the European Communities
European Colloquium – Proceedings.
Health effects of carbon monoxide. Environmental pollution.
1974. 433pp.
(d.e.f.i.n. EUR 5242).

Muller, K. H.

An air quality management system for an industrialized region
1975. 21pp.
(e. EUR 5290).

Literaturstudie über die ökonomischen Konsequenzun der Schäden und Belästigungen die durch die Luftverschmutzung durch Schwefeldioxid sowohl bei Materialien und der Vegetation als auch bei Mensch und Tier hervorgerufen werden
1974. 148pp.
(d. EUR 5134).

Nitrogen oxide emissions from energy generation within the EEC: 1970–1985
1974. 59pp.
(e. EUR 5136)

Air lead concentrations in the European Community
CEC Brussels
1973.
(d.e.f.i.n. EUR 4982).

Reducing pollution from selected energy transformation sources.
Chem. Systems Int. Ltd. Graham & Trotman, London. 229pp.
(e).

Critical review of the available physico-chemical material damage functions of air pollution.
1980. 98pp.
(e. EUR 6643)

Economic evaluation of damage to materials due to air pollution
1980. 97pp.
(e. EUR 664)

Elaboration of a common methodology for the biological surveillance of the air quality by the evaluation of the effects on plants
1980. 41pp.
(e. EUR 6842)

Elaboration of methods for determining the costs and benefits of implementing health protection standards concerning sulphur dioxide and suspended particulates
1980. 72pp.
(e. EUR 6852)

Exchange of information concerning atmospheric pollution by certain sulphur compounds and suspended particulates in the European Community: Annual report for January to December 1976
1980. 362pp.
(e. EUR 6472).

Study into the emission of air pollutants coming from the use of coal within the United Kingdom

1980. 246pp.
(e. EUR 6853)

Study on the impact of principal atmospheric pollutants on the vegetation
1980. 113pp.
(e. EUR 6644)

5. WATER POLLUTION

Studies on the radioactive contamination of the sea
M. Bernhard
Annual report
1964 (e)
(Association: European Atomic Energy Community – Euratom – Comitato Nazionale per l'Energia Nucleare – C.N.E.N.)

EUR 2543 e	1964 (1965)	35pp.
EUR 3275 e	1965 (1967)	30pp.
EUR 3635 e	1966 (1967)	30pp.
EUR 4244 e	1967 (1969)	66pp.
EUR 4508 e 1968/1969 (1971)		110pp.
EUR 4701 e	1970 (1971)	84pp.
EUR 4865 e	1971 (1972)	162pp.
EUR 5271	1972 (1974)	141pp.

Etude préliminaire sur la lutte contre la pollution dans le bassin rhénan
1973.
(f. EUR 5014 no. 3).

Etude de différentes pollutions constatées dans le bassin rhénan
1974. 136 pp.
(f. EUR 5133).

Les problèmes de pollution associés à la production de bioxyde de titane
1974. pag. diff.
(f. EUR 5195).

A study to determine the comparability of chemical analyses for drinking water quality within the European Communities
1976. 97pp.
(e. EUR 5542)

A European Community study on the determination of cyanides, phenols and hydrocarbons in surface water.
1975. 76pp.
(e. EUR 5377)

Water purification in the EEC
A state-of-the-art-review
Water Research Centre 467pp.
Pergamon Press 1977
(e).

Nuisances Thermiques.
Essai d'appreciation des dommages et de leurs palliatifs

1977. 200pp.
(e.f. EUR 5802).

Metallic Effluents of Industrial Origin in the Marine Environment
Prepared by the European Oceanic Association
1975. 230 pp.
Graham & Trotman, London 1977
and (e.f. EUR 5331).

Les rejets industriels contenant de l'arsenic en milieu marine
Assoc. Européenne Océanique
(f. EUR 5532).

Environmental Impacts and Policies for the EEC Tanning Industry
Prepared by Urwick Technology Management Ltd.
Graham & Trotman, London 1977. 90pp.

Cleaning and Conditioning Agents: their impact on the environment
Prepared by Environmental Resources Limited
Graham & Trotman, London 1978. 146pp.

Pollution by the Food Processing Industry in the EEC
Prepared by the Institut National de Recherche Chimique Appliquée
Graham & Trotman, London 1977. 193pp.

Biological aspects of freshwater pollution
1980. 224p
(e. EUR 6392)

6. POLLUTION BY WASTES

The problem of recycling waste oil in the Member States of the EEC
Hopmans
(f.e.d. EUR 5498).

Materials Flows in the Post-Consumer Waste Stream of the EEC
Prepared by H. C. Bailly and C. Tayart de Borms
Graham – Trotman, London 1977. 96 pp.

Secondary Materials in Domestic Refuse as Energy Sources
Prepared by Europool (Communication Consulting Research in Europe)
Graham & Trotman, London 1977. 92 pp.

Recovery of Metal Waste from Old Cars and Large Household Appliances
Graham & Trotman, London 1978. 96 pp.

Waste Materials Exchanges in the Member Countries of the EEC
Graham & Trotman, London 1978. 96 pp.

Disposal and Utilisation of Abattoir Waste in the EEC
Graham & Trotman, London 1978. 148 pp.

Product Planning
Graham & Trotman, London 1979. 311pp.

The Economics of Recycling
Graham & Trotman, London 1979. 167 pp.

The introduction of a tax with a view to reducing the consumption of non-renewable resources
1979. 119pp.
(e. EUR 6785)

Misure di radioattività ambientala: Ispra 1979
1979. 49pp.
(i. EUR 6785)

Demolition waste
1980. 190pp.
(e. EUR 6619)

Solid waste and chemical waste
1980. 853 pp.
(e. EUR 6596)

Summary of the study on measures taken by the Member States of the European Economic Community to improve the environment and to utilize mining wastes, with special reference to coal-mining and the lignite industry
1980. 25pp.
(e. EUR 6634)

Le traitement des déchets municipaux solides. Guide à l'usage des responsables locaux
1981. 95pp.
(e.)

7. NOISE

Damage and Annoyance caused by Noise
1975. 83pp.
(e. EUR 5398).

8. RESEARCH

Final reports on research sponsored under the First Environmental Research Programme (indirect action)
1978. 460pp.
(d.e.f. EUR 5970).

Second environmental research programme 1976–80: Reports on research sponsored under the first phase 1976–78
850pp.
(e. EUR 6388)

Second environmental research programme 1976–80: Reports on research sponsored under the second phase 1979–80
1110pp.
(e. EUR 7784)

Annex VI
European Parliament
Resolution of 20 November 1981

on the state of the Community environment

The European Parliament

— having regard to the second report submitted by the Commission of the European Communites to the Council on the state of the Community environment (1979 report),

— having regard to the Community's activities to date in the field of environmental protection which are reviewed in the communication from the Commission to the Council of 7 May 1980 (COM(80) 222 final),

— having regard to the report of the Committee on the Environment, Public Health and Consumer Protection and the opinion of the Legal Affairs Committee (Doc. 1–276/81),

FIRST PART: THIRD ACTION PROGRAMME ON THE ENVIRONMENT

1. General Remarks

A. The need for a third action programme

1. Considers that a third environment action programme is urgently needed;

2. Believes that a start should be made on this programme in 1981 and that it should as far as possible be based on the experience gained from earlier programmes and actions;

3. Reaffirms its support for the principles, goals and objectives of the first and second environment action programmes and for the measures contained therein;

B. Evaluation of previous programmes and actions

4. Proposes to this end that the Commission should report on the progress of completed programmes and explain where necessary why certain objectives were not attained in order that the third environmental programme can profit from this experience;

5. Urges the Commission also to consider the major national environmental programmes and evaluate their success;

6. (a) Welcomes the fact that the Commission has transformed its Environment and Consumer Protection Service into an independent directorate-general,
   (b) At the same time, considers it essential for the Commission to be provided with the necessary material resources and staff to carry out existing and future programmes;

C. Political principles and prerequisites

7. Welcomes the fact that environmental protection has developed from a purely defensive concept aimed simply at repairing damage into a preventive policy with the emphasis on precautionary measures;

8. Is of the opinion that this preventive policy should incorporate the following principles:

(a) the stand-still principle,
(b) the principle of the best technical means,
(c) the polluter-pays principle;

9. Is of the opinion that other areas of policy should be tied to ecological parameters with the aid of these principles and that the main instruments to be used for this purpose might be

(a) legislation on the environment,
(b) environmental impact assessment,
(c) environmental surveys,
(d) the laying down of standards;

10. Recommends that this idea of prevention be developed still further and that environmental protection be regarded as a positive determining factor and, as such,

(a) incorporated in long-term plans, economic scheduling and profitability calculations, and
(b) made an integral part of all relevant policies;

11. Proposes, therefore, that, in addition to the third programme, there should be an overall concept of, a strategy for, environmental protection with the individual measures envisaged forming part of this concept;

12. Recommends, therefore, since the desirable overall concept of environmental protection cannot be implemented immediately in its entirety, that the third programme should initially only deal with the most urgent problems;

13. Considers that there must be a clear statement from the Council now on the medium and long-term relationship between questions of energy, economic growth, and raw materials, on the one hand, and environmental protection on the other; the costs and benefits of the relevant policies must be investigated from this point of view;

14. Feels, also that an integrated, ecological frame of reference must first be prepared providing a clear definition of the 'polluter-pays' principle, the allocation of costs and the principle of cooperation;

15. Proposes that an account should be given of the effect, both beneficial and adverse, which environmental protection measures can have on employment;

16. Hopes that wherever possible the Commission will include a cost benefit analysis in all its future proposals in the environmental field;

17. Considers that the provisions of the Community legislation represent minimum provisions, i.e. that they should not affect the powers of Member States to apply or establish more stringent environmental provisions;

D. Progress to date

18. Recommends, further, that an assessment of what has already been accomplished in the environment sector should be appended, since the full range of individual measures so far taken is not universally known and the wealth of *ad hoc* projects does not give sufficiently clear indication of the efforts made and the resources deployed;

19. Considers also that national environmental protection measures should be included in this assessment, since this is the only way of providing an overall picture of what has been done and also of showing what results have been achieved, so that those areas can be identified in which more must be done or in which duplication of effort can be prevented;

II. Priorities for the third programme

20. Recommends that a list of priorities should be prepared for the third programme including in particular the following:

(a) development of cleaner alternative technologies and forms of energy,

(b) reduction of pollution from vehicles (noise and exhaust gases),

(c) measures to combat marine pollution,

(d) testing of effects of chemicals and drawing up of preventive measures,

(e) promotion of measures to eliminate waste products and to recover raw materials,

(f) reduction of water, land and atmospheric pollution,

(g) harmonization of the compulsory rules and provisions intended to adjust the conditions of competition,

(h) protection of nature and the countryside,

(i) research into the extent of soil pollution and the attendant pollution of the ground water and proposals to combat and prevent these types of pollution,

(j) consideration and aid for problems relating to nature and the environment in the Third World by means of the policies outlined in the World Conservation Strategy;

SECOND PART: MEDIUM AND LONG-TERM MEASURES (FUTURE ENVIRONMENTAL STRATEGY)

21. Proposes to the Commission that the following points should form part of a future overall concept, in so far as they cannot be incorporated into the third action programme:

I. Protection and rational management of natural resources

A. Pollution-free forms of energy and technology

22. Initiate policies to promote scientific and technological research with a view also to developing clean and alternative production processes and raw materials;

23. Encourage energy saving and the use of 'clean' energy;

24. In particular, place greater emphasis on research into environmentally clean methods of nuclear fusion and energy from hydrogen;

25. Take account of the results of those studies currently being done on the effects, advantages, disadvantages, quantities available and total costs—including expenditure on materials and equipment—of all types of energy;

26. Study the scope for producing energy from the biomass and examine in particular the potential for cultivation in those regions where farming is uneconomic and also consider, while assessing the cost involved, whether a change in the crops under cultivation could help reduce the surplus production of certain agricultural products;

B. Raw materials

27. Vigorously foster and encourage all projects designed to economize on or recycle raw materials or to develop and utilize substitute materials;

C. Waste disposal and recycling

28. Encourage and facilitate all projects involving the recovery of materials from waste products—in particular the collection of sorted domestic waste;

29. Ensure that more use is made of recycled paper, in particular in public administrations in the Member States;

30. Call for transfrontier cooperation in connection with the elimination of particularly dangerous waste products;

31. Examine how (particularly with regard to packaging) the volume of waste products could be reduced, for example by introducing tax incentives or constraints;

32. Devote greater attention to problems connected with sludge;

II. Reduction of environmental pollution

A. Water supplies and effluents

33. Arrange for careful investigations into future water requirements, and in particular into the volume of water available;

34. Adopt implementing Directives concerning water quality;

35. Harmonize national regulations on effluent disposal charges;

36. Adopt, at long last, a Directive on the discharge of list I substances into water, and establish for which substances regulations should be introduced as a matter of priority;

37. Provide for measures which can be used to combat not only specific but also scattered sources of water pollution;

B. The sea

38. Undertake research into marine pollution and introduce measures to eliminate and prevent further pollution;

39. Set up an adequate European surveillance body with the necessary powers to monitor and control shipping in Community waters;*

40. (a) Make the oil-tagging system, by which the oil in oil tankers can be made identifiable by adding metal particles to it, obligatory in European waters so as to make it easier to trace the origin of unlawful discharges,

   (b) Require the Member States to provide adequate storage facilities in all sea ports for the used oil which ships wish to discharge;

41. Take precautions to prevent any pollution from the exploitation of the ocean bed, in particular drilling for oil;†

42. Prevent and monitor the discharge of poisonous substances and pollutants on the high seas;

C. Air and atmosphere

43. Substantially lower the permitted levels of toxic exhaust gases from vehicles;

44. Take account of both direct and indirect aspects of sulphur dioxide (SO_2) pollution, particularly as regards its effect on forests, and fix acceptable sulphur levels for heavy heating oil‡ and coal intended for combustion;

45. (a) Draw up a scale of priorities to find the most effective means of combatting the photo-chemical pollution of air,

   (b) Define permitted levels for concentrations of NO_x, CH and oxidants and incorporate them in a Community Directive;

46. Continue with the preparation of common air quality standards and lay down, for specific pollutants, special standards applicable to areas surrounding the principal sources of emission;

47. Pay greater attention to problems connected with the pollution of the ozone layer with a more intensive study of possible effects on climate;

---

* See the resolutions adopted by Parliament on 16 January 1981 on the basis of the Maij-Weggen and Carossino reports (OJ No C 28, 9.2.1981, p. 52 et seq.).
† See the resolution adopted by Parliament on 16 January 1981 on the basis of the Maij-Weggen report (OJ No C 28, 9.2.1981, p. 56).
‡ This has already been done for light heating, oil; see Council Directive of 24 November 1975 (OJ No L 307, 27.11.1975).

D. Noise

48. Harmonize the noise emission levels of noisy products, particularly road vehicles, including goods imported from third countries;

49. Make a greater effort to combat noise at work (particularly from machines);

50. Take greater account, in the campaign against noise, of house and road construction;

51. Call upon

(a) the Council to approve without delay all the Directives now pending on the reduction of noise emission,

(b) the Commission to shorten the time limits laid down in the various Directives for fixing noise levels that are less harmful to human hearing;

E. Chemicals

52. Adopt the necessary legal provisions in implementation of the sixth amendment of the 1967 Directive,* in particular:

(a) establishing a list of existing substances and their characteristics;

(b) introducing a procedure for monitoring new substances;

(c) drawing up guidelines on classifying and labelling; in this connection, in view of the vast number of these substances and of the limited aids available, it is essential to:

find a procedure for establishing priorities, and

divide the work entailed among the Member States;

53. Urge the Council to adopt forthwith the Directive on the major accident hazards of certain industrial activities (Seveso Directive);†

F. Pesticides and fertilizers

54. Encourage

(a) the replacement of poisonous and non-degradable pesticides by products which cause minimal contamination of the atmosphere or waters,

(b) increased usage of fertilizers with lower nitrogen leachates;

55. Encourage research into and utilization of integrated pest control using natural and biological pesticides and life-forms;

56. Investigate to what extent biodynamic agriculture could be extended and promoted;

III. Protection and rational use of soil and land

57. Set up research and action programmes within the framework of European potential and competence to solve regional policy problems in

(a) rural areas,

(b) coastal and mountainous areas, and

(c) built-up areas (with particular reference to possible urban redevelopment schemes—slum clearance);

* OJ No L 259, 15.10.1979, p. 10 et seq.
† OJ No C 212, 24.8.1979, p. 4

58. Ensure that, above all in connection with town and transport planning greater importance is attributed to the rational and careful use of land and that clearance and redevelopment schemes take precedence over new developments on new sites;

59. Promote the conservation of nature and the landscape and ensure that

(a) nature reserves, areas of outstanding natural beauty and the natural course of rivers and streams are unspoiled,

(b) areas no longer under cultivation are revitalized, wetlands are preserved and restored and the consolidation of arable land, drainage systems and the development of large areas take greater account of the need to protect nature and the countryside, and

(c) in order to improve the ecological balance near towns, small-scale nature reserves or green-belt areas are created or natural habitats—of plants and animals—are preserved,

(d) representative chains of biogenetic reserves are created in the Community;

60. Take greater account of the impact of tourism (and recreational activities) on the countryside and minimize their harmful effects;

IV. Protection of animals and plants

61. Ensure that stronger emphasis is placed on the concept of animal protection in large-scale animal husbandry;

62. Prevent the use of chemicals to fatten animals;

63. Ensure that the number of animal experiments is reduced to the absolute minimum and review the legal provisions which, among other things, require experiments on animals for new pharmaceutical products;

64. Take steps

(a) to prevent the hunting of birds, whales and other endangered animals scrupulously observing the Community Directives and international conventions, and

(b) substantially to restrict seal hunting;

65. Assist with the conservation of areas which are important for animals and plants and restore polluted areas;

66. Draw up

(a) a Directive on plants in need of protection in the Community,

(b) a Directive on the protection of endangered vertebrates in the Community,

(c) a Directive on the protection of endangered invertebrates;

(d) a Directive to promote the harmonization of hunting legislation in the Community, in particular as regards hunting licence tests;

V. Environmental planning and administration

A. General points

67. Examine how universally valid bio-indicators for determining the quality of the environment might be established;

68. Provide ecological mapping of the EEC in a land register paying particular attention to border areas;

69. Establish binding criteria for environmental impact assessment and draw up guidelines for evaluating the implications of technology;

70. Compile easily interpreted statistics on the quality of the environment and ensure that the individual countries use a uniform method of presenting them;

71. Lay down obligations (compulsory reporting and other requirements) to be met by the relevant industries in order to prevent environmental disasters, and provide for reliability tests and controls which must, however, be designed to exclude the possibility of interference in management or other abuses;

72. Develop improved instruments for the early detection of hazards and for the reduction and reparation of damage, with particular reference to hazards and pollution with transfrontier implications;

B. Law and administration

73. Speed up the harmonization of national environmental law;

74. Arrange for a comparative study of environmental law and a compilation of national judicial rulings in this field;

75. Take steps as part of the process of harmonization of national environmental law, to ensure that legal provisions are drafted in a clear and unambiguous manner and that the lack of a clear political line does not lead to the application of too many vague legal concepts, which would place too great a strain on the courts;

76. Recommend, in connection with this harmonization process and as an essential aid in applying the law, the establishment of values, measurement and assessment standards and their constant adaptation to scientific and technical knowledge, as well as the extension in law of the scope allowed for administrative discretion, particularly in areas involving forward estimates and decisions on scientific or technical matters;

77. Recommend, as part of the administrative machinery, the setting-up of a control body within the authorizing authority with the necessary scientific and technical expertise;

78. Arrange for a comparative study on the most appropriate institutional structure for environmental protection in the Member States;

79. Adapt competition law so that products and processes which protect the environment are not placed at a disadvantage by cost distortions;

C. Transfrontier aspects

80. Develop suitable procedures for involving neighbouring countries in the planning and running of installations which could have transfrontier effects;

81. Develop adminstrative instruments to deal with transfrontier environmental pollution;

82. Devote special attention to the pollution of the Rhine, not only by pressing for action under the Convention for the protection of the Rhine against chemical pollution but also by urging stronger measures to be taken to deal with the problem of salt, taking into account Parliament's resolutions of 14 December 1979;[*]

[*] OJ No C 4, 7.1.980, p. 72 et seq.

83. Devote special attention to the Waddengebied, that important international nature area extending over 600 000 ha in the Netherlands, Germany and Denmark which, for a number of reasons (e.g. the Rhine), is threatened with destruction;

VI. Research

84. Insist that research should concentrate on the development of cleaner technologies and alternative energy sources and on the more economical of resources;

85. Investigate in detail and discover not only the damage caused by individual substances but in particular the cumulative effect of various nuisances and sources of pollution (especially in built-up areas), including combined and long-term effects;

86. Ensure that, in order to reduce costs, speed up developments and share intellectual and technological potential, environmental research is coordinated more effectively between the individual countries, universities and research agencies and between the industries concerned;

87. Enable each country to have constant access to the results of research in the individual partner countries;

VII. Information and education

88. Take steps to ensure that the exchange of information between countries is intensified and a flexible form of cooperation selected;

89. Seek to ensure that the general public is provided in good time with comprehensive information, in particular using audio-visual media, on the environmental situation in general and on measures which are proposed or have been implemented;

90. Propose more forcefully the inclusion of environmental policy among the subjects taught in schools;

91. Encourage the Member States to promote the training of environmental specialists both for educational establishments and for the public administration;

92. Inform in particular the European Parliament in good time of all projects and plans and, in this connection, to enable national projects to be included by extending the information agreement;

VIII: Economic policy aspects and assistance measures

93. Introduce, in connection with environment policy, a policy for the protection of resources which covers all matters relating to tax incentives and constraints and to investment aid;

94. Examine whether, and if so how, tax concessions could be granted for the development, utilization or manufacture of environmentally harmless products and substitute materials and for the recovery of materials from waste products;

95. Consider whether the prices of cheaper imported products which are harmful to the environment could be increased (corresponding to levies in the agricultural sector);

96. Seek agreement with the Third World countries to prevent the manufacture of environmentally harmful products from being transferred to their territory;

97. Seek to ensure that more positive account is taken in profitability calculations and cost-benefit analyses of the concept of environmental protection and the scarcity of resources;

IX. Ecology and development

98. Is of the opinion that, although nature and environment problems in the European Community are immense, they are even more serious in the countries of the Third World;

99. Is therefore convinced that the Community must pay much more attention to ecological problems in the Third World and that a separate chapter should be devoted to this subject in the third environment action plan;

100. Requests the Commission once again, following the resolution adopted by Parliament on 20 May 1980*, to pursue the objectives of the World Conservation Strategy which are based on

(a) the conservation of essential ecological processes and eco-systems,
(b) the maintenance of genetic diversity,
(c) the responsible use of species and eco-systems;

101. Is of the opinion that a permanent advisory committee on ecology and development should be set up to advise the Commission on the steps to be taken;

102. Is of the opinion that the ecological problems of the developing countries should become a permanent topic for consultations and the granting of aid within the framework of the Lomé Convention;

103. Is of the opinion that the activities of European financial institutions as well as those of the Community itself, where they affect aid granted to Third World countries, should, in cooperation with the recipient countries, be tied to ecological parameters with use being made in particular of environmental impact assessments and ecological checklists;

104. Is of the opinion that the Community should, in particular, offer a helping hand for:

(a) the conservation of the last tropical rain forests,
(b) programmes to prevent the spreading of desert,
(c) water management programmes,
(d) the promotion of small-scale and ecologically responsible agricultural systems,
(e) the promotion of small-scale, non-pollutant and energy-saving cooking and heating facilities,
(f) education about the environment;

105. Is of the opinion that products and production processes which are banned or considered undesirable in the Community on environmental or public health grounds should not be exported to Third World countries;

106. Is of the opinion that the measures outlined above should be implemented in close cooperation with UN bodies specialized in this field and with the Commission's Directorates-General for Development, Energy and Research and Economic and Financial affairs;

* OJ No C 147, 16.6.1980, p. 27.

THIRD PART: FINANCE—FUND FOR ENVIRONMENTAL MEASURES

107. Recommends that a fund for environmental measures—or equivalent facilities—be set up to finance, in particular,

(a) research and development in the field of clean energy sources and technologies,
(b) the protection of nature and the countryside,
(c) restorative action to compensate for particularly serious damage to the environment,
(d) research and development in the field of the economical use and recycling of raw materials,
(e) the dissemination of information on environmental protection, not least in schools;

108. Recommends that adequate appropriations be made available for this fund;

109. Instructs its President to forward this resolution and the report of its committee to the Council and the Commission.

Annex VII
Resolution Adopting the Third Action Programme (17.12.82)

RESOLUTION OF THE COUNCIL OF THE EUROPEAN COMMUNITIES AND OF THE REPRESENTATIVES OF THE GOVERNMENTS OF THE MEMBER STATES, MEETING WITHIN THE COUNCIL, ON THE CONTINUATION AND IMPLEMENTATION OF A EUROPEAN COMMUNITY POLICY AND ACTION PROGRAMME ON THE ENVIRONMENT

The Council of the European Communities and the Representatives of the Governments of the Member States, meeting within the Council, note that the projects to which the appended programme will give rise should in some cases be carried out at Community level and in others be carried out by the Member States.

With regard to the projects to be carried out by the Member States, the latter will supervise their proper execution, it being understood that for these projects the Council will exercise the co-ordinating powers laid down in the Treaties.

With regard to the projects in the programme to be carried out by the Institutions of the European Communities.

THE COUNCIL OF THE EUROPEAN COMMUNITIES,

Having regard to the Treaty establishing the European Coal and Steel Community,

Having regard to the Treaty establishing the European Economic Community,

Having regard to the Treaty establishing the European Atomic Energy Community,

Having regard to the draft from the Commission,

Having regard to the Opinion of the European Parliament,[*]

Having regard to the Opinion of the Economic and Social Committee,[†]

1. Whereas the Declaration of the Council of the European Communities and of the Representatives of the Governments of the Member States, meeting within the Council, of 22 November 1973[‡] calls for the implementation of a European Communities programme of action on the environment;

[*] OJ No C 182, 19.7.1982, page 102
[†] OJ No C 205, 9.8.1982, page 28
[‡] OJ No C 112, 20.12.1973, p. 1

2. Whereas the action programme was extended and amplified for the period 1977–1981 by the Resolution of the Council and the Representatives of the Governments of the Member States, meeting within the Council, of 17 May 1977;*

3. Whereas the tasks of the European Communities are laid down in the Treaties establishing the Communities;

4. Whereas in particular, in accordance with Article 2 of the Treaty establishing the European Economic Community, part of the latter's task is to promote throughout the Community a harmonious development of economic activities and a continuous and balanced expansion, which, even in changed economic circumstances, is inconceivable without making the most economic use possible of the natural resources offered by the environment and without improving the quality of life and the protection of the environment;

5. Whereas, consequently, the improvement of the quality of life and making the most economical use possible of the natural resources offered by the environment are among the fundamental tasks of the European Economic Community and whereas a Community environment policy would help accomplish this purpose;

6. Whereas the Council confirmed the objectives and principles of Community environment policy in its declaration of 27 November 1973 and reconfirmed them in its Resolution of 17 May 1977;

7. Whereas the programme of action on the environment of 22 November 1973, as extended and amplified on 17 May 1977, is still valid; whereas it should be updated, further implemented and supplemented for the period 1982–1986, by new tasks which prove to be necessary;

8. Whereas in particular, in addition to the projects already initiated (especially in the field of pollution reduction), making the most economic use possible of the natural resources offered by the environment requires the preventive side of the environment policy to be strengthened in the framework of an overall strategy and environment considerations to be integrated into other Community policies.

TAKES NOTE of the action programme annexed hereto and approves the general approach thereof;

DECLARES that it is particularly important for Community actions to be carried out in the following areas:

integration of the environmental dimension into other policies;

environmental impact assessment procedure;

reduction of pollution and nuisance if possible at source, in the context of an approach to prevent the transfer of pollution from one part of the environment to another, in the following areas:

● combatting atmospheric pollution, especially by NO_x, heavy metals and SO_2, *inter alia*, by implementing Directive 80/779/EEC of 15 July 1980 on air quality limit values and guide values for sulphur dioxide and suspended particulates;

● combatting fresh-water and marine pollution, *inter alia* by
  ● implementing Directive 76/464/EEC of 4 May 1976 on pollution caused by certain dangerous substances discharged into the aquatic environment of the Community and Directive 78/176/EEC of 20 February 1978 on waste from the titanium dioxide industry

* OJ No C 139, 13.6.1977, p. 1

- and the action programme of the European Communities of 26 June 1978 on the control and reduction of pollution caused by hydrocarbons discharged at sea;

- combatting pollution of the soil;

protection of the various aspects of the environmental aspects of the Mediterranean Region;

noise pollution and particularly noise pollution caused by means of transport;

combatting transfrontier pollution;

dangerous chemical substances and preparations; e.g. the supplementing and application of Directive 79/831/EEC amending for the sixth time Directive 67/548/EEC of 27 June 1967 on the approximation of the laws, regulations and administrative provisions relating to the classification, packaging and labelling of dangerous substances;

waste management, including treatment, recycling and re-use—toxic and dangerous waste, including transfrontier transport of such waste and the review of the list of toxic and dangerous substances and materials in the Annex to Directive 78/319/EEC of 20 March 1978;

encouraging the development of clean technology e.g. by improving the exchange of information between Member States;

protection of areas of importance to the Community which are particularly sensitive environmentally;

co-operation with developing countries on environmental matters;

NOTES with satisfaction that the Commission intends to be guided, as in the past, by the following considerations in drawing up its proposals:

the desirability of action at Community level;

the need to avoid any unnecessary duplication, by checking whether the subjects in question are already being dealt with satisfactorily by international bodies;

the need to assess, as far as possible, the costs and benefits of the action envisaged;

the need to take account of differing economic and ecological conditions and the differing structures in the Community;

the need to carry out careful research, analysis and consultation before proposals are submitted to the Council;

UNDERTAKES to act on these proposals wherever possible within nine months of the date on which they are submitted by the Commission or, as the case may be, of the date on which the Opinions of the European Parliament and of the Economic and Social Committee are submitted;

'STATES that the decision to make the financial resources necessary for implementing the Resolution and the action programme attached thereto will be taken in accordance with the usual procedures.'

*The numbers of the various paragraphs will not appear in the final version of the Resolution.